Local Control of Microvascular Perfusion

Colloquium Series on Integrated Systems Physiology: From Molecule to Function to Disease

Editors

D. Neil Granger, *Louisiana State University Health Sciences Center*

Joey P. Granger, *University of Mississippi Medical Center*

Physiology is a scientific discipline devoted to understanding the functions of the body. It addresses function at multiple levels, including molecular, cellular, organ, and system. An appreciation of the processes that occur at each level is necessary to understand function in health and the dysfunction associated with disease. Homeostasis and integration are fundamental principles of physiology that account for the relative constancy of organ processes and bodily function even in the face of substantial environmental changes. This constancy results from integrative, cooperative interactions of chemical and electrical signaling processes within and between cells, organs and systems. This eBook series on the broad field of physiology covers the major organ systems from an integrative perspective that addresses the molecular and cellular processes that contribute to homeostasis. Material on pathophysiology is also included throughout the eBooks. The state-of the-art treatises were produced by leading experts in the field of physiology. Each eBook includes stand-alone information and is intended to be of value to students, scientists, and clinicians in the biomedical sciences. Since physiological concepts are an ever-changing work-in-progress, each contributor will have the opportunity to make periodic updates of the covered material.

Published titles

(for future titles please see the website, www.morganclaypool.com/page/lifesci)

Local Control of Microvascular Perfusion
Michael A. Hill and Michael J. Davis
www.morganclaypool.com

ISBN: 9781615044368   paperback

ISBN: 9781615044375   ebook

DOI: 10.4199/C00061ED1V01Y201206ISP035

A Publication in the

COLLOQUIUM SERIES ON INTEGRATED SYSTEMS PHYSIOLOGY: FROM MOLECULE TO FUNCTION TO DISEASE

Lecture #35

Series Editors: D. Neil Granger, LSU Health Sciences Center, and Joey P. Granger, University of Mississippi Medical Center

Series ISSN

ISSN 2154-560X   print
ISSN 2154-5626   electronic

Local Control of Microvascular Perfusion

Dr. Michael Hill
Dalton Cardiovascular Research Center
Medical Pharmacology and Physiology
University of Missouri
Columbia, Missouri

Dr. Michael Davis
Medical Pharmacology and Physiology
University of Missouri
Columbia, Missouri

COLLOQUIUM SERIES ON INTEGRATED SYSTEMS PHYSIOLOGY:
FROM MOLECULE TO FUNCTION TO DISEASE #35

ABSTRACT

Local control of microvascular perfusion refers to the ability of individual tissues to maintain a relative constancy of hemodynamics in the face of changing perfusion pressure while meeting metabolic demands appropriately. The regulation of local blood flow, or *autoregulation*, and its underlying mechanisms have been a subject of considerable interest for over 100 years. Particular focus has been placed on the acute interaction of myogenic (pressure-induced) and metabolic (local production of vasodilator metabolites) parameters and how they interact with flow (shear stress)-dependent and conduction-based mechanisms to produce integrated local vascular network responses (for example, as seen during reactive and functional hyperemia). This monograph discusses each of these vasoregulatory phenomena while also considering evidence for their underlying cellular mechanisms. Further, an attempt is made to integrate the information into complex in vivo situations and consider their relevance to pathophysiological situations.

KEYWORDS

blood flow control; autoregulation; myogenic; metabolic; flow-mediated responses; conducted vasomotor responses; arterioles

Contents

Abbreviations

Abbreviation	Full Name
AII	Angiotensin II
BK_{Ca}	Ca^{2+}-activated, K^+ channels
BK_{Ca}	Large conductance Ca^{2+}-activated K^+ channels
Ca^{2+}_i	Intracellular Ca^{2+}
CPI-17	17 kDa PKC-potentiated protein phosphatase 1 inhibitor protein
DAG	Diaclyglycerol
ECM	Extracellular matrix
EDHF	Endothelial-derived hyperpolarizing factor
EGFR	Epidermal growth factor receptor
Em	Membrane potential
ERK	P42/44 mitogen-activated protein (MAP_ kinase)
IP3	Inositol-1,4,5-trisphosphate
K_v	Voltage-gated K^+ channels
MLC	Myosin light chain
Mypt1	Myosin phosphatase targeting subunit 1
NSCC	Non-selective cation channels
Pc	Capillary pressure
PKC	Protein kinase C
PMCA	Plasma membrane Ca^{2+} ATPase
SERCA	Sarcoplasmic endoplasmic reticulum Ca^{2+} ATPase
SMC	Smooth muscle cells
SR	Sarcoplasmic reticulum
VOCCs	Voltage-operated Ca^{2+} channels
VSM	Vascular smooth muscle

Acknowledgements

MAH and MJD are supported by grants from the National Heart, Lung and Blood Institute of NIH (HL092241, HL095486, and HL085119). Sincere thanks are extended to the members of our laboratories for their continued support and Drs. Meininger, Segal, Clifford and Braun for many valuable and insightful discussions.

CHAPTER 1

Introduction

The title of this volume, "Local Control of Microvascular Perfusion," refers to the ability of individual tissues to maintain a relative constancy of hemodynamics (local blood flow and pressure) in the face of changing perfusion pressure while meeting metabolic needs, including both the provision of nutrients and the removal of certain metabolites. Thus, during times of increased metabolic demand (such as occurs during muscular exercise), local microvascular regulatory mechanisms play a major role in providing the necessary increase in blood flow. Similarly, during an increase in perfusion pressure to a tissue, local regulatory mechanisms contribute to limiting hyperperfusion that would subsequently occur at the level of the exchange vessels or capillaries.

Importantly, the ability to regulate microvascular blood flow and pressure at the local level allows a degree of fine-tuning without necessarily involving systemic regulatory mechanisms (for example, by increasing cardiac output) that would impact all tissues. The principal mechanism underlying local control is the ability to appropriately alter the caliber of small arteries and arterioles, increasing vessel diameter by dilation or decreasing diameter by vasoconstriction. Alterations in vessel diameter are, in turn, critically dependent on the 'motor' or contractile activity of the smooth muscle cells located within the arterial wall. The smooth muscle cells, themselves, are further regulated by interactions with other elements of the vascular wall including the endothelium, the extracellular matrix and the three-dimensional architecture of the wall.

To accomplish local 'autoregulatory' control, the microvasculature exhibits several mechanisms that impact convective processes (in this context, blood flow), including myogenic constriction and dilation; metabolic dilation; and flow-induced modulation of vessel diameter and conducted vasomotor responses. These mechanisms will be discussed in terms of both their physiological importance and current understanding of the underlying cellular mechanisms. Further, the integration of these responses into network behavior is considered.

It is, however, important to realize that while these local regulatory mechanisms provide powerful control of hemodynamics at the tissue level, they do not exist in isolation. Interactions can occur with systemic-level control mechanisms, including those involving the sympathetic nervous and endocrine systems. The reader is referred to other volumes in this series, and publications such as the Handbook of Physiology, for more in-depth coverage of these topics. In addition, increasing evidence points toward considerable heterogeneity in the cellular mechanisms underlying local blood flow control between vascular beds. This further enables individual tissues to have a certain degree of autonomy in regulating local hemodynamics to meet tissue-specific needs. While this volume will mainly consider these local regulatory mechanisms from a general point of view, the reader is also encouraged to refer to volumes covering vasoregulation in specific vascular beds. For example, more specific information is provided in these volumes about regulation of the cerebral microcirculation through neurovascular coupling [66] and regarding regulation of the renal microcirculation via pre- and post-glomerular vascular mechanisms and tubuloglomerular feedback [330].

A further consideration is that the microvasculature is not a static structure. While the following chapters will largely focus on acute (seconds to minutes) mechanisms of vasoregulation, on a longer time frame local blood flow can be impacted by a number of vascular remodeling events that may lead to changes in cellular function, modification of vessel wall structure or even the degree of vascularity (including both vasculogenesis and rarefaction) [266, 267, 292]. Such changes may occur in response to changing environmental conditions (for example, changes in altitude and exercise training); pathophysiological states (for example, hypertension, diabetes mellitus and cardiac dysfunction); and developmental stages including aging and pregnancy.

$\cdot$ $\cdot$ $\cdot$ $\cdot$

CHAPTER 2

Historical Perspectives

From a historical point of view initial interest in the microcirculation began with the availability of the microscope. This critical development allowed Malpighi and Leeuwenhoek to directly observe previously invisible vascular pathways connecting arteries to veins (refer to Hwa and Aird [191] for an historical discussion of this period). This allowed red cells and their movement to be visualized in transparent tissues and direct confirmation of Harvey's theory that blood circulated from the arterial to the venous side of the circulation perhaps via capillaries.

Studies specifically examining the local control of blood flow have a rich history dating back to the beginning of the 20th century and earlier. While the concept of autoregulation of blood flow was introduced in 1931 in studies of the renal circulation, the earlier experiments of Bayliss [20], in 1902, related to the myogenic properties of blood vessels, had suggested that tissues exhibited an ability to regulate flow in the face of changing perfusion pressure (Figure 1). As highlighted by Johnson in 1986 [204], Bayliss wrote "the peripheral powers of reaction possessed by the arteries is of such a nature as to provide so far as possible for the maintenance of a constant flow of blood through the tissues supplied by them, whatever may be the height of the blood pressure, except so far as they are directly overruled by impulses from the central nervous system." These observations of Bayliss laid largely unexplored for some 40 years as other investigators had suggested his data could be explained entirely by the effects of pressure/flow changes on the availability of vasodilator metabolites [401]. Subsequent studies did, however, demonstrate and characterize the autoregulatory ability of the renal, cerebral, coronary, splanchnic and skeletal muscle circulations [for examples, see 26, 133, 134, 347, 350]. Importantly, Folkow in the late 1940s and early 1950s showed that denervated preparations developed pressure-dependent vascular tone [134] and that autoregulation of blood flow was partly explained by non-neural, pressure-dependent mechanisms [136]. These publications stimulated substantial interest in understanding the relative contributions of metabolic and myogenic regulation of blood flow and paved the way for future studies of the underlying cellular mechanisms. The proceedings of the symposium 'Autoregulation of Blood Flow,' published in *Circulation Research* in 1964 (Volume XIV/XV, Supplement 1), provide a resource describing many of the early studies in this field.

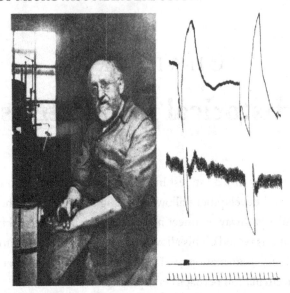

FIGURE 1: Sir William Maddox Bayliss, often credited for the first description of the arterial myogenic response. The tracings to the right show original records examining the effect of compression of the aorta on hind limb volume. Taken from *Handbook of Physiology, 1980* with permission.

In the 1960s and 1970s, the introduction of in vivo preparations and video microscopy (building on earlier use of cine-photography, for example, [31, 68, 408]) for quantitative study of the microcirculation, enabled direct study of microvascular reactivity at the level of individual tissues [11, 12, 210, 406, 407]. This further allowed direct observation of network behavior in conjunction with measurements of local intravascular pressures [137, 194, 195, 352, 418], vessel diameters [196, 197] and red blood cell velocity [38, 147, 406] (with subsequent calculations of flow). During this period, there were many novel technical developments that enabled specific questions to be addressed—for example, the ability to study coronary hemodynamics in the beating heart [298] and for local intraluminal pressure to be manipulated without affecting systemic hemodynamics or the pressure gradient for blood flow [83, 277, 417]. Later, the development of isolated vessel myographs [293] and the ability to isolate and cannulate single arterioles [106] allowed pressure-induced vasoconstriction to be studied at the single vessel level. Importantly, these approaches allowed small vessels to be studied in the absence of the influence of parenchymal tissues and under conditions in which mechanical forces and chemical environment could be tightly controlled. The availability of fluorescent indicators (for example, Ca^{2+} sensitive indicators including fura 2 [279] and fluo 4 [299]), relatively specific pharmacological inhibitors, sharp electrode recordings [165] and microbiochemical approaches (including, western blotting, quantitative PCR and manipulation of protein expression [35, 59, 322, 382, 414] subsequently enabled more direct studies of the underlying cellular signaling mechanisms. More recently, many of these mechanistic intact-vessel studies have been supported by single cell studies using sophisticated approaches such as patch clamping [86] and atomic force microscopy [378]. Combined with the increasing availability of transgenic and disease-state models, our understanding of the mechanisms underlying the myogenic properties of small arteries has not only been improved but has also been broadened.

Many additional important studies have, of course, added to our understanding of mechanisms by which local blood flow control is affected. For example, as early as 1890, Roy and Sherrington were describing the vasoactivity of neural, metabolic and pharmacological stimuli on the blood supply of the brain [333]. Starling [21, 370] stressed the importance of both hydrostatic and oncotic pressures in the function of the microcirculation. Further, the works of Krogh [233] some 90 years ago and Schretzenmayr [344] in 1933 illustrated the phenomenon of conducted vasomotor responses (through observing vessel dilation remote from the site of stimulus) which was later shown by Hilton [183] to be independent of neural input, setting the stage for direct characterization at the level of the microcirculation, and mechanistic studies underlying intercellular communication. Krogh [233] further proposed that blood flow must be linked to tissue oxygenation, arguing that if this was not the case, then supply of oxygen must be markedly excessive at rest or inadequate during periods of maximal demand.

While it is not practical to present a full historical picture of all studies contributing to our knowledge of local control of the circulation, further illustrations are given where appropriate in the following Chapters. For additional information, the reader is referred to the *Handbook of Physiology* volumes, relating to the microcirculation, published in 1980 [2] and in revised form in 2008 [1]. An additional resource is found in the Microcirculation Video Archive, which is held at the University of California San Diego library.

· · · ·

CHAPTER 3

Microcirculatory Elements Affecting Blood Flow Control

Before focusing on the function of arterioles it is first instructive to define the term 'microcirculation.' The microcirculation generally refers to the small blood vessels within a tissue and is comprised of a network of small arteries/arterioles, a capillary network and the venules/small veins (Figure 2). Another element often included is the network of lymphatic capillaries and collecting ducts, which in many cases can be observed to parallel the blood vessels. In general, the arterioles branch from small arteries distributing blood flow to discrete regions of the tissue where the capillaries allow exchange of nutrients and the uptake and convective washout of products of metabolism. In addition

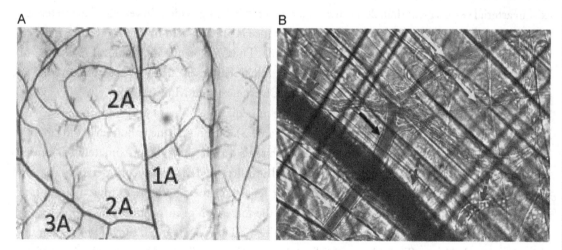

FIGURE 2: (A) Polymer cast of the arteriolar vasculature of rat skeletal muscle. The tissue is supplied by the first order arteriole (1A; diameter approximately 120 μm), which branches into smaller 2A and 3A vessels. Note that in this technique the polymer does not fill the capillaries or venules but is restricted to the resistance vessels or arterioles. (B) Microcirculation in skeletal muscle as visualized using in vivo microscopy. Arterioles (solid yellow arrows) and a venule (dotted blue arrow) can be seen on a background of skeletal muscle. Image was collected using transillumination video microscopy and a high resolution CCD camera. Image is courtesy of Dr. Steven Segal, University of Missouri.

to this, and as already emphasized, the arterioles, by actively regulating their diameters, play a major role in regulating local blood flow and pressure. Similarly, capillaries perform additional functions contributing to fluid balance between the vasculature and the tissues, while venules play major roles in permeability and inflammatory responses as well as in returning blood to the systemic veins.

In the following sections, we will specifically consider arterioles with emphasis on both structural make-up and their function as a resistive element within the microcirculation.

3.1   DEFINING AN ARTERIOLE—STRUCTURAL AND FUNCTIONAL CONSIDERATIONS

As a general rule, the control of microcirculatory blood flow lies at the level of the arterioles. This ability arises from the high level of intrinsic basal resistance, or contraction, in these vessels that can be reduced or further increased by neurohumoral and mechanical stimuli. An alternate description for this basal level of contraction is 'tone.' While there is currently no precise definition of an arteriole, such vessels are typically less than approximately 150 μm in diameter [87] and demonstrate inherent tone. Exceptions, however, exist as larger-diameter cerebral [19] and coronary [63] vessels, as well as a number of 'feed arteries' (term used to define the smallest artery directly feeding a tissue; typically used in reference to skeletal muscle vasculature), often exhibit a considerable level of intrinsic tone. Chilian et al. reported approximately 25% of basal coronary vascular resistance to exist in arterial vessels great than 200 μm with a further 20% being evident in vessels of diameter between 100 and 200 μm [63]. Importantly, this resistance is functional in nature and can be reduced by acute administration of the vasodilator, papaverine. Thus, in considering the local regulation of blood flow, these observations have led to a largely functional description of 'resistance arteries' as referring to all arterial vessels possessing tone (see Christensen and Mulvany [65]).

From a structural perspective, arterioles have a media consisting of 1–2 smooth muscle cell layers, lined by a single layer of endothelial cells (Figure 3). Separating the endothelial and smooth muscle cells is the internal elastic lamina (IEL) which is a thin 'sheet' (approximately 0.3 μm in thickness) largely comprised of the protein elastin. In reality, the IEL shows considerable variation from being an intact sheet with periodic holes (or fenestrae) to a more discontinuous layer (see Figure 10). Holes in the IEL are believed to allow the formation of myoendothelial gap junctions (MEGJs) that act as communication pathways for electrical and/or chemical signaling between the intimal and medial layers. The abluminal side of the vessel is typically lined by a complex layer of extracellular matrix proteins, often containing a number of cell types. More specifically, the adventitia contains various extracellular matrix proteins (including collagen, elastin, laminin, fibronectin) and additional cellular elements including pericytes, fibroblasts and nerve bodies and terminals. Accumulating evidence also suggests that some vessels may also have a type of pacemaker cell [168, 319], analogous to the interstitial cells of Cajal seen in the intestinal wall [337, 404]. Brief descriptions of each of the components of the arteriolar wall are given later in this chapter.

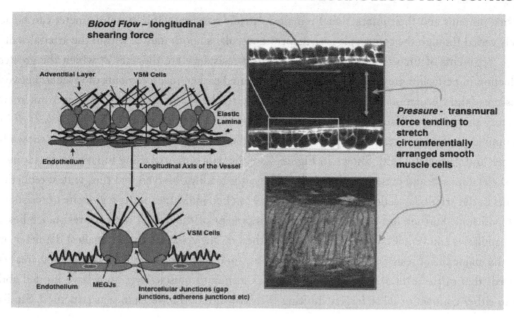

FIGURE 3: The arteriolar wall consists of circumferentially arranged smooth muscle cells and an inner layer of endothelial cells lining the vessel lumen. Communication between these cells occurs via gap junctional contacts as well as through interactions with extracellular matrix proteins. The figure illustrates this arrangement in schematic form (left) and in sections of cannulated and fluorescently labeled arterioles (right). On the right, the upper figure is a cross-section through the wall and lumen, while the lower figure shows the orientation of the smooth muscle cells. The figure further illustrates the major forces exerted on the vessel wall—transmural pressure and shearing force. Figure is taken from reference [181].

It is important to realize that the components of the artery wall do not exist independently. Interactions between the elements of the arteriolar wall play important roles in both the modulation of contractile function and the maintenance/adaptation of vascular structure. Such interactions are facilitated by both the vessel architecture (for example matrix elements traversing the adventitia and media and cellular junctions) and through release of vasoactive factors in a locally acting, paracrine, fashion.

3.2   ARTERIOLES AS SITES OF VASCULAR RESISTANCE

Vascular resistance refers to the force that must be overcome to move blood through the circulation. The determinants of resistance include vessel diameter, their length and the viscosity of the blood. Under physiological conditions, vessel length and viscosity (determined largely by the red blood cells, plasma proteins and their interactions) remain relatively constant while vessel diameter can be actively varied through the contractile activity of the vascular smooth muscle within the arterial wall.

Segments of the vasculature with the greatest resistance are the sites at which the greatest reduction in perfusion pressure is observed. As indicated earlier, measurements of vascular network pressures (and calculations of resistance) were facilitated by the development of in vivo microcirculatory preparations. Using direct glass micropipette measurements (*servo-null technique*) of intraluminal pressure, the relationship between vessel diameter and perfusion pressure was shown by Fronek and Zweifach [139]. Shown in Figure 4 are the data obtained using in vivo preparations of cat skeletal muscle and mesentery. From these data, it is readily observed that the greatest reduction occurs in the arteriolar sections of the vascular network bed indicative of the major site of resistance to blood flow. Further, only a relatively small component of the overall network resistance lies in the capillaries and venules. Capillaries, although they are the vessels with the smallest diameter, are not the major site of resistance because their number (and therefore, overall cross-sectional area) far exceeds that of the arterioles. Further, as they lack a smooth muscle layer, they have limited ability to either contract or relax, largely limiting their contribution to resistance as structural. Similar

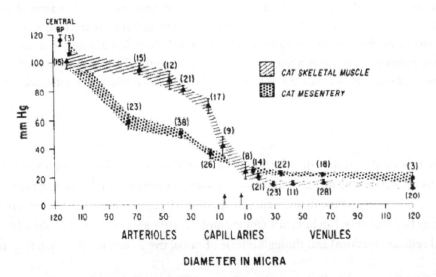

FIGURE 4: Intraluminal pressure distribution relative to vessel diameter in cat skeletal muscle and intestinal mesentery. Central blood pressure (BP) is shown as a reference. The figure highlights the following: (1) the decline in pressure that occurs at the level of the precapillary resistance vessels and (2) that some differences exist between tissues in regard to the distribution of vascular resistance. From reference [139].

According to Poiseuille's Law, flow (Q) can be expressed in terms of the pressure gradient (ΔP), vessel radius (r) and length (l), and the fluid viscosity (η):

$$Q = \frac{\Delta P \pi r^4}{8 l \eta} \qquad (1)$$

Analogous to Ohm's Law, Q = pressure difference/resistance which allows equation 1 to be rearranged to illustrate the determinants of resistance (R)

$$R = \frac{8 l \eta}{\pi r^4} \qquad (2)$$

FIGURE 5: Poiseuille's Law relating flow to the pressure gradient and fourth power of the vessel radius (1). Equation (2) shows the relationship between resistance to flow and the inverse of the radius to the fourth power. Both equations highlight the impact made by the radius such that small changes in this variable will lead to large changes in flow and resistance.

relationships between intraluminal pressure and vessel diameter were subsequently demonstrated in a number of vascular beds and species [19, 33, 63, 89, 97, 148, 276, 376].

Fronek and Zwiefach [139] further showed that vasodilation (with papaverine) decreased the resting levels of resistance and pressure. The pressure–diameter relationship when vessels are fully dilated (maximally relaxed) represents the passive state. The difference between the passive state and that under resting conditions can be referred to as functionally accessible resistance (as opposed to that explained by structural considerations). This effect of vasodilators was subsequently demonstrated in a number of other tissues [17, 63].

Also evident from Figure 4 is that the pressure distribution differs somewhat between the skeletal muscle and intestinal tissues. A greater resistance is evident in the smaller arterioles of resting skeletal muscle (compared to mesentery) perhaps reflecting their suitability to respond to locally produced metabolic stimuli.

From a quantitative perspective, the relationship between blood flow and vessel radius can be appreciated by considering Poiseuille's Law (Figure 5, Equation 1). Flow (Q) is proportional to the pressure gradient and the radius (r: to the fourth power) while being inversely proportional to the blood viscosity. As the impact of the radius is raised to the fourth power, it is evident that small changes in this parameter dominate the equation. Using a variation of Ohm's Law ($Q = \Delta P/R$; $R =$ resistance) allows rearrangement of Poiseuille's Law to show that the resistance is inversely proportional to the fourth power of the radius (Figure 5, Equation 2).

3.3   SMOOTH MUSCLE

Arteriolar smooth muscle cells are oriented approximately perpendicularly to the long axis of the vessel [154, 403] (Figure 3). Thus, on shortening or contraction, the vessel lumen is reduced, while on relaxation, the lumen dilates. Depending on the diameter of the arteriole, more than one smooth muscle cell (SMC) may overlap to form the circumference, although in small arterioles (such as shown for precapillary arterioles in ureter and vas deferens), a single SMC may wrap around the vessel more than once [39]. Arteriolar smooth muscle typically acts in a unitary fashion, relying on intercellular connections or gap junctions for the spread of stimuli between cells.

The role of SMCs in the acute local control of perfusion directly relates to their ability to rapidly undergo active contraction or relaxation thus affecting arteriolar diameter. For detailed information on the regulation of smooth muscle contraction per se, the reader is referred to the volume in this series entitled "Regulation of Vascular Smooth Muscle Function" by Khalil [215] and recent review articles [74, 361]. In brief, smooth muscle contraction is initiated by electromechanical or pharmacomechanical coupling [362]. The former refers to stimuli directly initiating a change in membrane potential (Em) (for example, KCl, stretch) while the latter largely refers to receptor-mediated stimuli that lead to the generation of second messengers (including IP_3 and DAG). In many cases, however, smooth muscle contraction is initiated via a combination of these mechanisms. Particularly relevant to arterioles, mechanical stimulation may be a reflection of this combined/dual activation [87, 88, 346].

Pharmacomechanical stimulation is initiated by the binding of a vasoactive substance to its cell surface receptor (see Table 1 for examples of receptor-mediated activation of VSM). Binding to the receptor activates trimeric G-protein signaling which in turn stimulates phospholipases (most often phospholipase C) to generate second messengers including IP_3 and diacylglycerol (DAG). IP_3 binds receptors (typically the IP_3R1 isoform [152, 295]) on the SR to release Ca^{2+} to the cytosol, while DAG is an activator of a number of isoforms of PKC and several ion channel proteins (for example, the non-selective cation channel protein, TrpC 6).

Following these events, an increase in global intracellular Ca^{2+} often occurs both through the effects of IP_3 on the SR and entry from the extracellular space via L-type voltage-operated Ca^{2+} channels (VOCCs). The opening of the VOCCs is facilitated by the initial depolarization of the plasma membrane. The mobilized Ca^{2+}_i binds to calmodulin (at a stoichiometry of 4 molecules of Ca^{2+} to a single calmodulin molecule), which then activates myosin light chain kinase (MLCK). Activated MLCK subsequently catalyzes the phosphorylation of the 20 kDa myosin regulatory light chain (at serine 19) which causes a conformational change in the myosin head group, enabling a physical acto-myosin interaction and cross bridge cycling (Figure 6) [74]. These dynamic protein–protein interactions thus determine the level of contraction.

TABLE 1: Pharmacomechanical coupling in arteriolar smooth muscle			
VASOACTIVE SUBSTANCE	RECEPTOR*	SIGNAL TRANSDUCTION MECHANISMS	ACTION
Adrenergic			
	α1	Gq, PLC, ↑IP$_3$, ↑Ca^{2+}, VOCCs	constriction proliferation
	α2	Gi, cAMP, PLC, ↑IP$_3$, Ca^{2+}, VOCCs	constriction
	β1	Gs, ↑cAMP, ↓Ca^{2+}, K$_{ATP}$	relaxation
	β2	Gs, ↑cAMP, ↓Ca^{2++}, K$_{ATP}$	relaxation
Angiotensin II	AT1	PLC, ↑IP$_3$, ↑Ca^{2+}, PKC, ERK, EGFR	constriction
			growth/hypertrophy
Endothelin	ET$_A$	PLC, ↑IP$_3$, ↑Ca^{2+}, PKC, ERK	constriction proliferation
Vasopressin	V1	Gq, PLC, ↑IP$_3$, ↑Ca^{2+}, PKC	constriction
Eicosanoids			
	IP	↑cAMP, ↓Ca^{2+}, K$_{ATP}$	relaxation
	TP	PLC, ↑IP$_3$, ↑Ca^{2+}	constriction
5-HT	2A	PLC, ↑IP$_3$, ↑Ca^{2+}, VOCCs	constriction
Adenosine	A1, A2	Gi, Gs, ↑cAMP, ↓Ca^{2+}, K$_{ATP}$	relaxation

*Table 1 provides examples only and is limited to agents acting directly on VSMCs and their receptors. For more extensive details relating to receptor subtypes, details of signal transduction and actions in the microcirculation, see reference [405].

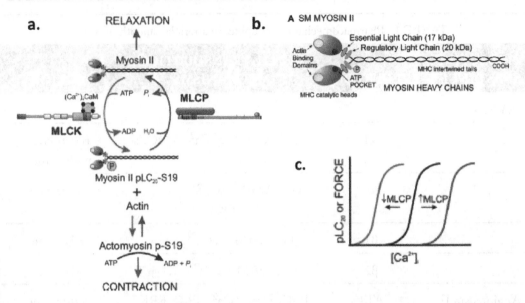

FIGURE 6: Role of actin and myosin in the regulation of vascular smooth muscle contraction and relaxation. (A) The interaction of the contractile proteins actin and myosin is regulated by the balance of myosin light chain kinase (MLCK) and myosin phosphatase (MLCP) activities. (B) Smooth muscle myosin (SM myosin II) consists of two heavy chains and two catalytic myosin head regions that bind actin. Ca^{2+}-dependent phosphorylation of the 20-kDa regulatory light chain is required to allow binding to actin. (C) Modulation of MLCP activity leads to changes in the apparent Ca^{2+} sensitivity of contraction. From Cole and Welsh [74].

Contraction via Ca^{2+}–calmodulin activation of MLCK is supported by modulation of MLC phosphatase activity (Figures 6 and 7). This process has been referred to as 'Ca^{2+} sensitization' as inhibition of the phosphatase activity results in a decreased rate of dephosphorylation of MLC_{20} prolonging acto-myosin interaction and enhancing contraction [74, 361]. This can also be illustrated by a leftward shift in the relationship between intracellular $[Ca^{2+}]$ and generated force, such that a relatively greater contraction would occur than predicted for a given change in $[Ca^{2+}]_i$ (Figure 6). Inhibition of myosin phosphatase is thought to occur by two predominant mechanisms, both of which have been shown to be activated by a variety of vasoactive stimuli. Specifically, through RhoA (a small molecular weight G-protein)-dependent activation of Rho kinase, or via activation of protein kinase C (PKC). Rho kinase-mediated phosphorylation of the myosin phosphatase targeting subunit 1 (MYPT1) at threonines 633 and 855 prevents docking of the phosphatase with myosin, thus preventing dephosphorylation of MLC_{20} and maintaining contraction. Similarly,

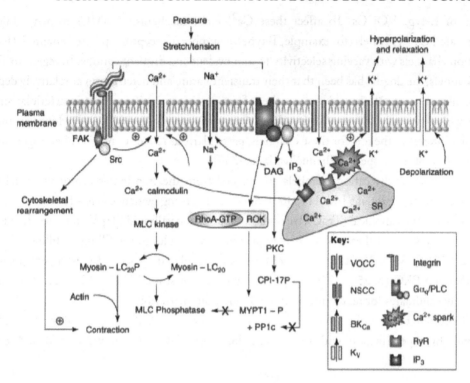

FIGURE 7: Schematic diagram illustrating specific signaling mechanisms currently thought to mediate arteriolar myogenic vasoconstriction. Emphasis has been placed on pathways for which detailed biochemical and/or electrophysiological evidence are available, other pathways are inferred from evidence obtained in other smooth muscles and await direct confirmation in arterioles. See text for further details; from Hill et al. [180].

PKC can phosphorylate CPI-17 (17 kDa PKC-potentiated protein phosphatase 1 inhibitor protein), which then acts as an inhibitor of the myosin phosphatase.

In addition to the mechanisms mediated by myosin light chain kinase and phosphatase (referred to as thick filament-mediated regulation), contraction may also be supported by thin filament (or actin-based) regulatory mechanisms and/or rearrangement of the SMC cytoskeleton [159, 218]. Thin filament regulation may involve proteins including caldesmon and calponin, the activities of which are modulated by phosphorylation [150, 218, 289, 359].

While the above has stressed Ca^{2+} entry via L-type VOCCs, it is important to consider the involvement of other cation entry mechanisms. These additional mechanisms may directly supply Ca^{2+} or cause membrane depolarization (often by influx of Na^+) that subsequently leads to the opening of L-type VOCCs. To affect these Ca^{2+} entry mechanisms, SMCs display additional voltage-gated Ca^{2+} channels (for example, T-type), a number of receptor-operated channels (ROCs) and cation channels with varying selectivity to monovalent and divalent cations. In regard to T-type Ca^{2+} channels, the dogma has been that their transient nature and inactivation at relatively depolarized membrane potentials would prevent them from playing a significant role in arteriolar smooth muscle which typically shows a resting Em of approximately –40 mV [300]. However, there is increasing interest in the role for these channels, particularly as relates to regional heterogeneity and the expression of tissue specific channel variants [235].

A number of the cation channels are formed from the transient receptor potential (Trp) family of membrane proteins [67, 111, 400]. Specifically, strong evidence exists for important contributions from the canonical (TrpC), melastatin (TrpM) and vallinoid (TrpV) families. In addition to cation entry, some of these channels may play important roles in the filling/re-filling of the SR with Ca^{2+} [316]. Several of these channel types are modulated by second messengers (including Ca^{2+}, DG and PKC [6, 250, 357] and perhaps mechanical stimuli [111–113, 175, 414], making them likely candidates for regulating contractile function in arterioles.

Relaxation of arteriolar VSM cells occurs via a decrease in Ca^{2+}_i often resulting from plasma membrane hyperpolarization, with subsequent closure of VOCCs (Figures 7 and 8), and removal

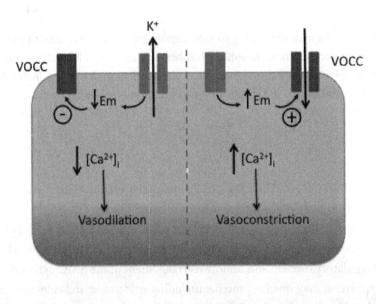

FIGURE 8: Effect of K^+ channels on Em, intracellular Ca^{2+} and vascular tone. Under physiological conditions, opening of K^+ channels promotes K^+ efflux causing membrane hyperpolarization. The more negative Em results in decreased opening of VOCCs, lowering of $[Ca^{2+}]_i$ and SMC relaxation. In contrast, closure of K^+ channels depolarizes the cell membrane increasing the open probability of VOCCs and contraction. See references [201, 300, 360].

of cytosolic Ca^{2+} via extrusion (via the plasma membrane Ca^{2+} ATPase; PMCA) or re-uptake into the SR (via the sarcoplasmic endoplasmic reticulum Ca^{2+} ATPase; SERCA). In addition, Ca^{2+} desensitization may occur as a result of increased myosin phosphatase activity and an increased rate of myosin light chain dephosphorylation [318, 433]. Hyperpolarization typically occurs as a result of the opening of K^+ channels, in particular, voltage-gated K^+ channels (K_v) and large conductance, Ca^{2+}-activated K^+ channels (BK_{Ca}). The K^+ channels can be opened as a negative feedback mechanism for vasoconstrictors (via depolarization and increased $Ca^{2+}{}_i$) or by vasodilators (generating NO, cyclic nucleotides).

3.4   ENDOTHELIUM

The importance of the endothelium in regulating vasomotor function has been firmly established over the past 30 years. Rapid advancements during this period have led to the realization that the endothelium is far more than an inert barrier separating the blood from the underlying muscle and led to the awarding of two Nobel Prizes directly related to endothelial function. These developments are detailed in a number of recent reviews [114, 129, 141].

As in larger vessels, endothelial cells line the arteriolar lumen as a monolayer. Their long axis is typically along the length of the vessel such that they are aligned in the direction of flow. Endothelial cells are extensively coupled to one another through gap junctions (containing connexins, Cx37, Cx40, Cx43 [94]) and to the underlying smooth muscle layer via myoendothelial gap junctions (MEGJs). The MEGJs traverse the internal elastic lamina (IEL) that separates endothelial and smooth muscle cells through discrete holes (examples can be seen in Figure 10). Coupling of the endothelium is a key component in the 'spreading' of vasomotor mechanisms including propagated dilation and endothelial-dependent hyperpolarization (see Chapter 6).

From a functional point of view, it is well accepted that the endothelium is a major source of vasoactive paracrine factors—both vasodilators and vasoconstrictors. The vasodilators include nitric oxide (NO), metabolites of arachidonic acid (including prostacyclin, epoxyeicosatrienoic acids) and endothelial-derived hyperpolarizing factors (EDHF). The EDHFs are not yet fully defined and likely include actual factors (including metabolites of arachidonic acid and H_2O_2) as well as the spread of hyperpolarization through the low-resistance gap junctions [114, 141, 338, 341]. Thus, it is important to realize that endothelial-dependent hyperpolarization (EDH) is a point of convergence for a number of pathways that lead to vasodilation. The endothelium not only produces factors leading to dilation but also produces a number of constrictor agents. Considerable interest has been shown in their possible role in pathophysiological states. Endothelial-derived vasoconstrictors include endothelin and metabolites of arachidonic acid (including 20-HETE) [131, 399].

Production of endothelial-derived vasoactive factors occurs in response to both receptor-mediated and mechanical stimuli. Many of these stimuli activate signaling mechanisms that converge on an increase in endothelial cell $Ca^{2+}{}_i$ that in turn contributes to the activation of enzymes, including endothelial nitric oxide synthase (eNOS) and cyclooxygenase (key enzymes for the production of NO and prostaglandins, respectively).

As with SMCs, both receptors and ion channels [130, 225, 303] play important roles in regulating endothelial cell function and subsequently the tone of arterioles. Included among recep-

	TABLE 2: K$^+$ channels* in VSM and endothelial cells			
K$^+$ CHANNEL	ALTERNATE NAMES	G* (APPROX)	MODES OF REGULATION	SITES OF EXPRESSION
K$_V$	Voltage-gated K$^+$ channel	5–30 pS	Voltage, protein kinases	VSM
K$_{ir}$	Inward rectifier		Extracellular K$^+$, voltage	VSM
K$_{ATP}$	ATP-sensitive K$^+$ Channel	15–50 pS	Availability of ATP/ADP, protein kinases	VSM
BK$_{Ca}$	–Large conductance, Ca^{2+} activated, K$^+$ channel. –Slo	250 pS	Ca^{2+}, voltage, protein kinases, steroid hormones, lipids	Predominantly VSM. Some evidence for expression in microvascular ECs.
SK	–Small conductance, Ca^{2+} activated, K$^+$ channel. –K$_{Ca}$2.3	2–20 pS	Ca^{2+} via constitutively bound calmodulin	EC
IK	–Intermediate conductance, Ca^{2+} activated, K$^+$ channel. –SK4, K$_{Ca}$3.1	20–60 pS	Ca^{2+} via constitutively bound calmodulin	EC/proliferating VSM

*G = Conductance (note — values are approximate as these represent families of channels and values are dependent on recording conditions).

**For further details of nomenclature and properties, see [7, 75, 161, 201, 247, 300, 409].

tors expressed by arterial endothelial cells are those for acetylcholine, bradykinin, substance P, ATP, 5HT, angiotensin II and endothelin. In regard to ion channels, endothelial cells express several K^+ channels (Table 2), with small and intermediate conductance Ca^{2+}-activated K^+ channels being particularly important for components of endothelial-dependent vasodilation.

3.5   EXTRACELLULAR MATRIX COMPONENTS—THE IEL AND THE ADVENTITIA

It has long been recognized that the extracellular matrix components of the arteriolar wall contribute a structural element providing a framework for the positioning of the cells that comprise the intima and the media. The endothelium is separated from the smooth muscle cells of the media by the internal elastic lamina (IEL). As referred to earlier, the IEL is composed mainly of the protein, elastin. The IEL, while forming a layer between endothelial and smooth muscle cells, is perforated

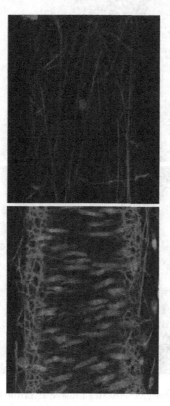

FIGURE 9: Small mesenteric artery adventitial elastin fibers. Fibers are stained with Alexa 633 hydrazide (red) and nuclei with Yo-Pro iodide (green) [71]. Images were taken using a confocal microscope and reconstructed in 3D using Imaris software. Ella, Clifford, Stupica, Meininger and Hill (unpublished).

by holes or fenestrae. These holes seem to show considerable variation in size between vascular beds and perhaps vessel sizes, causing the IEL to vary in appearance from a sheet with very discrete, punctate, holes to one that more closely resembles a meshwork (Figure 10). The IEL, and in particular the holes, appears to perform an important functional role in endothelial—smooth muscle cell communication. In this regard, the holes appear to correspond with myoendothelial junctions that are physical projections formed by plasma membrane elements of the two cell types. The projections are a site of concentration for gap junctions that form a type of intercellular channel (see also Chapter 7). These junctions have been proposed to allow the spread of electrical signals (current) or small signaling molecules (for example, IP_3) between the cell layers [114, 141, 341, 363]. Alternatively, the holes have been proposed to facilitate the diffusion of paracrine factors between the two cell layers.

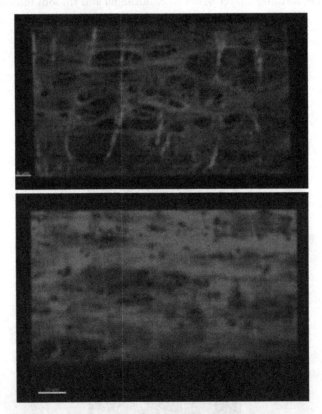

FIGURE 10: Example IEL layers from a cremaster muscle arteriole (upper) and a cerebral artery (lower). IEL was stained with Alexa 633 hydrazide. Images were taken using a confocal microscope and reconstructed in 3D using Imaris software. Note the more mesh-like nature of the IEL in the cremaster vessel compared to the more discrete, or punctate, appearance of the cerebral vessel IEL. Ella, Clifford, Stupica, Martinez-Lemus, Meininger and Hill (unpublished).

Directly underlying the endothelial cells is a matrix layer referred to as the basement membrane. This layer exists as a thin sheet, or lamina, and is comprised principally of type IV collagen, laminin, fibronectin and heparin sulfate proteoglycans. As a general property, the basement membrane provides an adherence point for endothelial cells through cell–matrix adherence proteins (including integrins). In specialized structures, such as renal glomeruli, the basement also contributes to the filtration barrier. In vasculogenesis, basement membranes form between endothelial cell tubes and pericytes and contribute to the stabilization of the developing vessel [375].

The outer layer of the vessel, or adventitia, contains a number of extracellular proteins including collagen (predominantly the fibrillar Types I and III) and elastin. The elastin fibers may act to bear stress in the longitudinal direction, particularly in vessels that are subjected to axial stretch (for example in skeletal muscles and intestinal mesentery) [71]. In many small arteries, the outer longitudinal elastin fibers appear to branch, sending smaller fibers into the vessel wall which in turn may interact with smooth muscle cells via matrix binding receptors such as integrins and elastin binding proteins. Fibrillar collagens aggregate into bands that 'criss-cross' the vessel in a pattern referred to as a 'Chinese finger trap' [145]. Both elastin and collagen are highly branched proteins with elastin being more highly branched than collagen. The differences in composition and character of elastin and collagen contribute to their differences in functional behavior. For example, in terms of their contribution to the passive elastic properties of an artery, elastin exerts its influences at low intraluminal pressures while collagen impacts at higher pressures.

3.6   ADDITIONAL CELLULAR ELEMENTS WITHIN THE VASCULAR WALL AND ADVENTITIA

In addition to vascular smooth muscle and endothelial cells, several other cell types are associated with the vascular wall and may impact contractile function. These include neural cells that utilize a variety of vasoactive transmitters (in addition to those of the sympathetic nervous system) of peptidergic (for example, CGRP, substance P, neuropeptide Y), purinergic (for example, ATP) and gaseous (for example, NO) nature. While the presence of these neurons and their transmitters are recognized, any specific role in the local control of blood flow is less well appreciated.

Recent evidence has suggested that the arterial wall may contain cells specialized to act as pacemakers [34]. Candidate cells have been identified both in large arteries and veins (portal) as well as in relatively small (rat middle cerebral and guinea pig mesenteric) arteries. While not fully characterized, such pacemaker cells may be a specialized population of smooth muscle cells [169, 319] and have been estimated to constitute approximately 5% of the cellular component of the vessel wall. The cells have been described as being, themselves, non-contractile with irregular-shaped cell bodies from which a number of thin processes extend. This morphology has led to the suggestion that they may resemble interstitial cells of Cajal (pacemaker cells within the intestinal wall

[336]) although they do not express the marker c-kit when first isolated from the vessel wall. The idea of a vascular wall pacemaker is intriguing as such a cell type could conceivably contribute to coordination of responses/activities across the vascular wall or rhythmic phenomena such as vasomotion (although other models exist to explain this rhythmic phenomenon [162] and not all the vessels in which these pacemaker cells have been identified can be considered to demonstrate a classic rhythmic behavior). In support of such a role, such cells isolated from portal vein show rhythmic changes in intracellular Ca^{2+} and membrane depolarization [168]. As differences in smooth muscle cell phenotype within the normal vascular wall have previously been described [312], it is also possible that these cells represent a cell line destined for functions unrelated to contractile activity (for example, in growth-related or repair responses).

Adventital fibroblasts appear to perform multiple tasks including contributions to both physiological [98, 262] and pathophysiological (inflammation, fibrosis, remodeling) states [72, 262]. Relevant to the regulation of arterial tone fibroblasts have been shown to produce and release vasoactive factors such as angiotensin II and endothelin-I although these factors may also modulate extracellular matrix production and vessel structure. Heterogeneity in fibroblast populations (as assessed by cell surface markers and their possible pluripotency) further complicates our understanding of their exact roles.

The adventitia also contains mast cells and macrophages that have been implicated in inflammatory responses. During such conditions, these cells produce NO and reactive oxygen species which could impact local vascular tone. Mast cells have also been suggested to mediate constriction in response to adenosine [103]. While adenosine is a powerful locally acting vasodilator in many systems (via A1/2 receptors), the mast cell constrictor response occurs via adenosine's action on an A3 receptor. Adenosine causes the mast cell to degranulate releasing locally acting factors such as histamine. Supporting this mechanism, a known mast cell degranulating agent (compound 48/80) mimicked the effects of adenosine [103].

Pericytes are an actin-containing cell that have been identified in the walls of small blood vessels particularly capillaries and post-capillary venules. While implicated in vessel growth (including angiogenesis both through their pluripotency and secretion of factors that regulate the process), their contractile properties and sensitivity to metabolic products such as CO_2 [58] and lactate [424] have raised the possibility that they may also contribute to the regulation of capillary blood flow. Further, they interact with endothelial cells [374, 375] to secrete matrix components of the basal lamina (basement membrane), which in the brain contributes to the blood brain barrier [102]. Interestingly, pericytes form junctional complexes (including gap junctions) with endothelial cells suggesting that they may communicate by transfer of signaling molecules and via mechanical forces.

Although not strictly associated with the adventitia, it has also been suggested that periadventitial adipose tissue releases locally acting vasoactive substances [146]. These factors (in a manner analogous to endothelial-derived factors) may diffuse to the vascular smooth muscle cells and elicit direct vasomotor responses. Evidence has been provided that these yet to be fully characterized vasoactive substances act via an effect on voltage-gated K^+ channels. The activity can be distinguished from a number of factors released by adipocytes (including leptin and adiponectin) and is independent of the endothelium. Candidate relaxing factors suggested to date include the gaseous mediator, H_2S [146].

Recent studies have identified (using both GC/mass spectrometry and vessel myography approaches) one perivascular adipose tissue-derived relaxing factor as palmitic acid methyl ester, or PAME. Interestingly, this substance exerts a smooth muscle relaxing effect through Kv and inhibits angiotensin II production by the fat cells making it a candidate for involvement in hypertension [248].

Although the study of the effect of periadventitial fat on the vasculature has largely been conducted on conduit vessels, a number of small arteries are in close contact with fat. Further, many factors released from fat, including adiponectin and cytokines, are known to impact vasomotor tone [155] although the precise mechanisms are yet to be fully characterized. Thus, this presents an interesting area for a potential novel paracrine mechanism for the local regulation of the vasculature.

Collectively, the available information on the function of the myriad of cell types in the adventitia and perivascular tissues suggests that this compartment should not be ignored in terms of it providing vasoactive paracrine factors. Conceivably, this allows an outside-in signaling mechanism (analogous to an inside, or endothelial, directed mechanism) for communicating signals across the vessel wall [72]. At present, evidence does not exist for its specific role in local blood flow regulatory mechanisms under physiological conditions; however, an indirect effect through modulation of blood vessel tone and a contribution to pathophysiological disturbances of blood flow is likely.

•   •   •   •

CHAPTER 4

Autoregulation

4.1   BLOOD FLOW AUTOREGULATION

Autoregulation refers to the ability of tissues, under conditions of constant metabolic requirement, to maintain a relatively constant blood flow despite changes in perfusion pressure. Schematic representations of this phenomenon are shown in Figures 11 and 12. The figure further contrasts the response of an active autoregulating vessel with that of either a rigid tube or a tube with distensible walls but lacking the ability to actively contract or relax. When perfusion pressure falls, after an initial passive collapse, arterioles (and some small arteries) dilate, lowering vascular resistance and increasing blood flow towards baseline levels of blood flow. Alternatively, when pressure increases arterioles constrict, increasing vascular resistance and again returning blood flow towards basal levels.

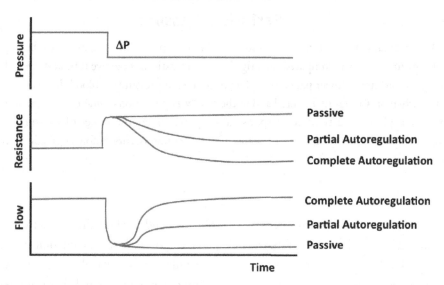

FIGURE 11: Schematic diagram illustrating the temporal characteristics of autoregulation and the relationships between pressure, vascular resistance and blood flow. The schematic applies to any vascular bed showing no (passive), partial or near complete autoregulation. Note the slight increase in resistance, evident in the passive state, when pressure decreases reflects passive collapse of the vessel. Note also that y axes are not to scale.

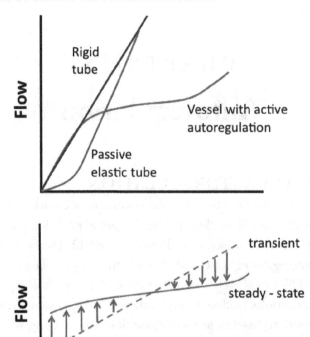

Perfusion Pressure

FIGURE 12: Schematic diagrams further illustrating the concept of autoregulation. In the upper panel, the effect of pressure on flow is compared for a rigid tube, an elastic, but passive tube and a vessel showing active autoregulation. In the lower panel, the effect of perfusion pressure on blood flow is shown for the transient state when pressure is first altered and in the steady-state response where autoregulatory adjustments have taken place. The transient response also approximates the effect of pressure on a non-autoregulating vessel. Typically, a delay of some 30–90 seconds occurs before a steady-state flow is attained.

The cellular mechanisms (myogenic, metabolic and endothelial) that collectively underlie autoregulation are the subject of the following chapters. Importantly, autoregulation is an intrinsic property of the local vascular network occurring independently from (while interacting with) the sympathetic nervous system. Thus, Granger and Guyton demonstrated whole body autoregulation (approximately 75% compensation following step changes in pressure of 25 to 50 mm Hg) after destruction of the central nervous system consistent with both a non-neural mechanism and a contribution from the majority of tissues [151].

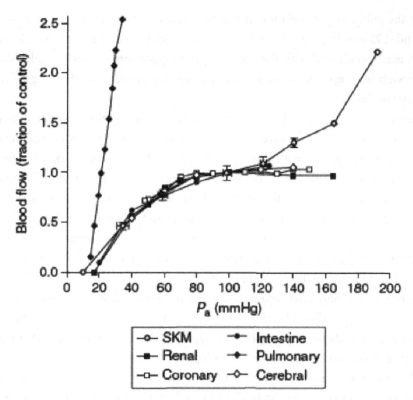

FIGURE 13: Whole-organ pressure–flow relationships. Reproduced from reference [88] where data re-calculated from the studies described in references [40, 220, 241, 304, 367]. Flows have been normalized to the flow at $Pa = 100$ mm Hg, except for pulmonary data which are normalized to P pulmonary artery = 21 mm Hg. SKM, skeletal muscle. Figure is taken from reference [88].

The physiological significance of autoregulation can be appreciated in a number of situations. For example, when pressure to an organ is decreased due to a narrowing of its supply vessels auto-regulation can return, within certain limits, blood flow towards normal. Similarly, autoregulation can contribute to the overall adjustment in hemodynamics following postural changes—for ex-ample when an arm is moved from above to below heart level [172]. During systemic hypotension, autoregulatory responses in cerebral and coronary circulations allow flow to be maintained to vital organs while sympathetic vasoconstriction reduces flow to other organs.

Autoregulation is observed at varying degrees in most tissues throughout the body. Figure 13 illustrates steady-state pressure–blood flow relationships for 6 differing vascular beds. With the exception of the pulmonary vasculature, it can be noted that between perfusion pressures of approximately 60–130 mm Hg, there is a relative constancy of blood—this reflects an 'autoregulatory range' which corresponds well with the normal operating pressure of these vessels. Note that the pulmonary vessels both operate at much lower perfusion pressures and show a passive pressure–blood flow relationship.

Not shown on the diagram, skin and cutaneous tissues (one of the largest organs of the body) of humans also do not show substantial autoregulatory responses as they are largely controlled by sympathetic innervation in line with thermoregulatory needs. In addition, skin has relatively low metabolic requirements for which basal blood flow exceeds and in certain regions is complicated by the presence of arteriovenous anastomoses. Thus, it can be argued that if the cutaneous circulation exhibited autoregulation (in line with metabolic requirements), this would oppose its thermoregulatory role. However, autoregulatory responses are evident in cutaneous tissues of some species such as in the bat wing [91] and spontaneous myogenic constriction can be observed in vitro in small arteries taken from human skin biopsies [125, 126]. Evidence of in vivo skin blood flow autoregulation can also be seen but only after neural blockade and substantial reduction in perfusion pressure (for example, following ganglionic blockade) [110, 172]. Therefore, it is likely that these vessels do exhibit inherent autoregulatory capacity but it is weak relative to a number of other tissues and is overridden by other control systems under physiological conditions.

Another organ showing distinct autoregulatory properties is the liver. The liver receives blood via both the hepatic artery (approximately 25% of total hepatic blood flow) and the portal vein (approximately 75% of total). Autoregulation is observed in the hepatic arterial circulation but not via the portal vein. Presumably this relates to the metabolic demands of the liver being predominantly met by the oxygenated hepatic artery supply as opposed to the deoxygenated blood entering via the portal vein.

Before considering underlying mechanisms in subsequent chapters, it is instructive to first consider some of the background whole organ and in vivo microcirculation studies that have provided strong support for the concept and existence of autoregulation. The first reports demonstrating autoregulatory behavior date back to the 1930s in studies of the renal circulation [170] although Bayliss had earlier noted that the kidneys were particularly responsive to changes in tension [20]. In the following 20 years, an increasing number of studies demonstrated the autoregulatory capacity of the cardiac, cerebral and splanchnic circulations. However, a role for autoregulation was not supported by all investigators. Interestingly, the intensity of interest in this field led to a group of diverse investigators meeting in 1962 to not only discuss the concept (as well as hypotheses for underlying mechanisms) but to openly demonstrate their preparations [209].

Whole organ and tissue experiments were performed using either auto-perfused approaches or isolated organs or tissues that were typically pump perfused. In some of these studies cannulation of side branches or advancement of fine plastic cannulae to distal sites of the vasculature were used to determine sites of vascular resistance. Collectively, these approaches established the concept of autoregulation and implicated the site of control to be located distal to the level of small arteries.

The subsequent availability of in vivo microcirculatory preparations, which could be studied using video microscopy, allowed autoregulatory events to be studied at the single vessel level and to determine the exact sites at which changes in resistance occur. Quantification was provided by measurements of vessel diameter, red cell velocity, microvascular pressures and vessel number. From these measurements, parameters such as volumetric blood flow and capillary density could be calculated. Importantly, these approaches supported the findings of whole organ studies with respect to the occurrence of autoregulation and clearly demonstrated the important contribution of arterioles.

Using these methods, Johnson and Wayland [210] demonstrated that during a decrease in perfusion pressure capillary red blood cell velocity did not change in direct proportion with the change in pressure. Often red cell velocity remained constant indicating that the vascular network was not simply behaving in a passive manner. Morff and Granger [288] later showed, using an exteriorized cremaster microcirculatory preparation in combination with graded occlusion of the aorta, that while on average arterioles showed evidence of autoregulation, individual responses varied from decreased through maintained flow to actually increased (super-regulated) flow. This observation stresses the heterogeneity of responsiveness [379] and also the importance of considering the integrated response of the overall network. Subsequently, a number of studies confirmed that smaller arterioles tended to be the most reactive (both in terms of rate and magnitude of response), reinforcing the importance of this section of the vasculature in autoregulatory responses (Figure 14).

Using exteriorized microcirculatory preparations, several investigators addressed the question of whether the prevailing metabolic state, and specifically the level of oxygenation, affects blood flow autoregulation. These studies have been important to our understanding of whether autoregulation can be explained by myogenic (see Chapter 5) or metabolic (Chapter 6) mechanisms.

Direct visualization of the microcirculation has also demonstrated additional phenomena that would impact autoregulatory responses. In particular, an increase in the number of flowing capillaries, or capillary recruitment, has been observed during vasodilator responses. Recruitment of capillaries both lowers vascular resistance and increases the surface area for nutrient/metabolite exchange between the vasculature and the surrounding parenchymal tissues.

Any active contribution of venules and capillaries to changes in vascular resistance during blood flow autoregulation appears to be minimal. During reduction of perfusion pressure, these vessels either maintain their diameter or passively collapse [29, 189]. Conceivably passive collapse could increase resistance on the venular side of the circulation and oppose autoregulation. However,

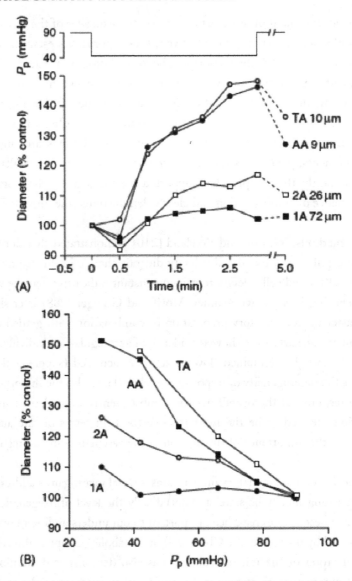

FIGURE 14: Effect of arteriolar size, or branch order, on the time course or extent of dilation in response to a decrease in perfusion pressure from 90 to 40 mm Hg. Figure is taken from reference [88] and has been redrawn from the original data of Davis [82]. Experiments were conducted in the bat wing and pressure manipulations performed using the 'pressure box technique' [82]. AA, arcuate arteriole; TA, transverse arteriole.

the changes in venular/capillary diameter are small and may be offset by factors such as the number of capillaries with flow.

4.2   AUTOREGULATION OF CAPILLARY PRESSURE

Whether or not capillary pressure is, itself, autoregulated has been more controversial than the concept of blood flow autoregulation. Maintenance of capillary pressure during fluctuations in overall perfusion pressure is an attractive concept as this may prevent unwanted fluid shifts within the capillary bed. Extending this, Loutzenhiser has argued that in regard to glomerular capillaries, regulation of capillary pressure is of prime importance with its main function being to prevent glomerular hyperfiltration and pressure-induced damage [257]. The glomerular capillaries may, however, be a special case as these exchange vessels operate at a considerably higher pressure (approximately 60 mm Hg) compared to most regional capillary beds (approximately 25 mm Hg).

As with blood flow autoregulation, data relating to the regulation of capillary pressure has been obtained using whole organ and in vivo microcirculatory approaches. On the basis of in vivo studies performed on cat intestine and dog hindlimb circulations, Johnson [206] suggested that changes in arterial inflow or venous outflow pressure produced changes in arterial resistance that would serve to minimize changes in capillary hydrostatic pressure. Mellander and colleagues [30, 203] demonstrated that tissue volume of cat hindlimb skeletal muscle was nearly constant over a wide range of systemic arterial pressures (30 to 170 mm Hg). Assuming that a constant tissue volume was consistent with constant Pc, these authors concluded that "autoregulation of Pc" occurred via myogenic adjustments of arteriolar tone. This conclusion assumes, however, that other Starling forces are not involved in control of tissue volume and that other local regulatory mechanisms did not contribute significantly to the changes in vascular resistance [33, 83, 149]. Even though these whole organ techniques have significant limitations (for example, that during venous pressure elevation there are also changes in flow, shear stress and metabolite accumulation [90]), direct measurements of Pc in microcirculatory preparations have, to date, been unable to completely resolve this issue.

Direct measurements of Pc with glass micropipettes have produced conflicting results as to whether capillary pressure is tightly regulated during changes in Pa. Gore and Bohlen, in studies of rat and cat intestinal preparations, reported Pc to fall in line with reductions in Pa. Zwiefach, however, found Pc in cat mesenteric preparations to be very well regulated over an arterial pressure range of 90–170 mm Hg. Using the bat wing preparation together with the pressure box techniques Davis found Pc to be partially regulated as Pa was varied. The advantage of the pressure box technique is that it provides a non-invasive method for altering perfusion pressure thus not altering neurohumoral regulatory mechanisms. In addition, the bat wing preparation does not require surgery or anesthesia.

Thus, it still remains uncertain as to whether, as a general property, that Pc is actively regulated or whether its apparent regulation occurs as a result of blood flow regulation. A number of factors contribute to the difficulty in resolving this question. For example, in determining Pc whole organ methods provide an average measurement, while in direct micropipette measurements, there may be a selection bias in that larger vessels with more obvious blood flow are often selected for study (capacity for autoregulatory adjustment may be limited in such vessels). Further, Pc is also likely to be determined, in part, by venular resistance that is affected by network structure (for example the presence of arcades) and during decreases in Pc, the rheological properties of blood (including local hematocrit, viscosity and red cell aggregation).

•   •   •   •

CHAPTER 5

Myogenic Vasoconstriction and Dilation

5.1   PHYSIOLOGICAL SIGNIFICANCE

Myogenic, or pressure-induced, vasoconstriction provides a level of basal vascular tone in arterioles upon which neurohumoral stimuli can act to either lower vascular resistance or further increase constriction [85, 87, 88]. Thus, this basal level of constriction places arterioles in a position to rapidly respond to changes in blood flow demand. Often the myogenic properties of arterioles are subdivided and referred to in terms of 'myogenic response' (the vasomotor response to an acute change in intraluminal pressure) and steady-state 'myogenic tone' [274]. Alternatively, Osol and Gokina [308] have proposed a three-phase model for myogenic constriction based on the initial development of myogenic tone at low intraluminal pressures, a maintained phase of relatively stable myogenic tone across a physiological range of pressures, and finally a loss of myogenic tone (forced distension) at high intraluminal pressures. Whether these facets of myogenic responsiveness are mechanistically distinct is currently an area of considerable interest.

In addition to providing a level of basal tone, the arteriolar myogenic response is thought to be a major mechanism underlying autoregulation. When perfusion pressure increases, under conditions of unchanged metabolic need, myogenic constriction will attenuate any increase in flow distal to the point of constriction. Similarly, if perfusion pressure decreases, myogenic dilation will tend to maintain blood flow. While it was argued that a metabolic mechanism (see [134, 207, 350, 401]) could explain the vasomotor response, the demonstration of myogenic reactivity in in vivo preparations, where the pressure gradient for blood flow was not changed [277], and in single arterioles cannulated in vitro (Figure 15) confirms the existence of a myogenic mechanism [434].

Myogenic constriction has also been implicated in the local regulation of capillary pressure (Pc). As an increase in perfusion pressure would shift the Starling forces towards filtration, and ultimately cause edema, pressure-induced constriction of arterioles would limit the transfer of pressure to the exchange section of the vasculature. Thus, this is an example of autoregulatory behavior and analogous to the effects of myogenic constriction on local control of blood flow.

If Pc is regulated by myogenic mechanisms, it would appear that this is more effective in the case of pressure increases where myogenic constriction would limit fluid loss from the downstream

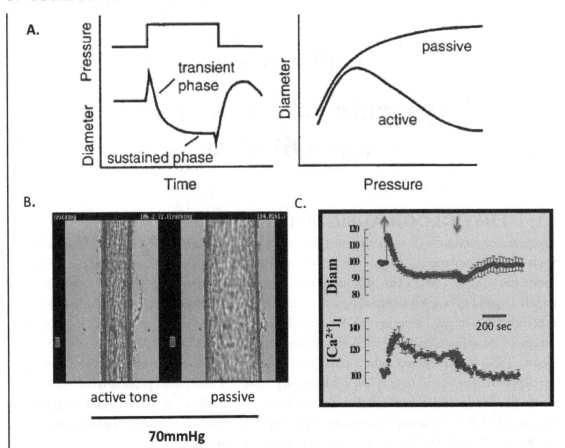

FIGURE 15: Panel A shows schematic representations of arteriolar myogenic behavior. The left panel A illustrates the temporal changes in diameter to an acute increase in intraluminal pressure. At the onset of the pressure pulse the vessel passively distends before contracting to a steady-state. On return of the pressure to its starting level, the vessel first passively collapses before dilating to its original diameter. The right panel compares the diameter of a vessel under passive and active conditions over a range of intraluminal pressures. Note at low pressures the 'active' vessel behaves passively before reaching a threshold pressure for myogenic activation. Panel B shows images of an isolated cannulated cremaster arteriole under myogenically active conditions compared to the passive state. Under active conditions, the vessel shows constriction to approximately 50% of its passive diameter. In both situations, the vessel was held at an intraluminal pressure of 70 mm Hg in the absence of flow. Panel C shows group diameter and intracellular Ca^{2+} data for cannulated cremaster muscle arterioles ($n = 5$; in the absence of flow) subjected to an acute 50–120 mm Hg pressure step (up arrow). Return of pressure to 50 mm Hg (down arrow) shows that the response is fully reversible. From Zou et al. [434].

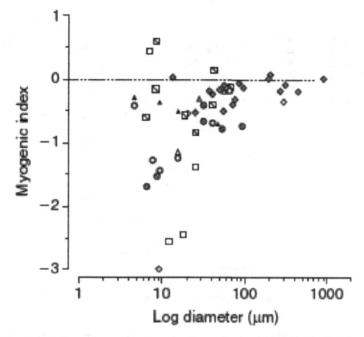

FIGURE 16: Myogenic index (change in diameter per unit change in pressure) in arterioles of varying diameter. As vessel diameter decreases, there is an increase in myogenic reactivity as shown by the calculated myogenic index. Original data were taken from studies of hamster cheek pouch arterioles [84] as redrawn in reference [88].

capillaries. Consistent with this, studies in bat wing illustrated that over a similar pressure range Pc was more tightly regulated during controlled pressure increases as compared to pressure decreases [82, 83]. As mentioned earlier, in the renal circulation Loutzenhiser has argued that myogenic constriction of the afferent arteriole is a major factor in protecting the glomerular capillaries against pressure-induced damage [257]. Glomerular capillaries do, however, operate at considerably higher intraluminal pressures than do capillaries of the general circulation (approximately 45–60 mm Hg (measured directly in rat [41] while indirectly determined in humans and dog [219, 384]) as compared to approximately 25 mm Hg in capillaries of other tissues).

As mentioned above, arteriolar myogenic constriction has been demonstrated both in vivo and in vitro. The latter approach has most often used isolated cannulated arteriole preparations under isobaric conditions and in the absence of flow through the lumen. Figure 15 shows the temporal nature of the response of a group of skeletal muscle arterioles (from rat cremaster muscle) to an acute pressure increase from 50 to 120 mm Hg and on return to the starting level of pressure. When

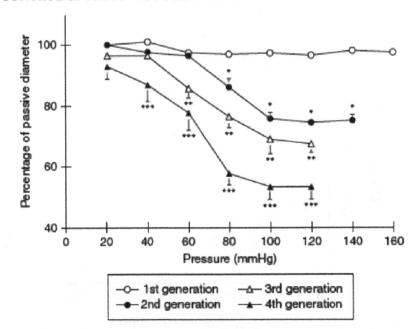

FIGURE 17: Myogenic reactivity of arterioles of differing branching order. Data are shown for four successive branch orders of rat mesenteric arteries and show increasing pressure-dependent constriction in the smaller branches. From reference [377].

pressure is first increased the vessel passively distends after which active vasoconstriction occurs. The response reverses on decreasing intraluminal pressure, with an initial passive collapse followed by an active dilation. Of note, in many arterioles the steady-state level of contraction achieved after an acute increase in pressure results in a diameter significantly smaller than the starting vessel diameter. Not all arterioles, however, constrict below control level rather maintaining diameter or distending slightly. Such responses are often still considered myogenic if the response to pressure is significantly different to that of a purely passive vessel (where contractile is inactivated by application of dilators and/or removal of extracellular Ca^{2+}). Examples of myogenic behavior are schematically illustrated in Figure 15.

The myogenic reactivity of an arteriole/small artery can vary both between tissues and along a particular network. Studies of arterioles from hamster cheek pouch [84] and rat intestinal mesentery [377] show that the extent of pressure-induced myogenic constriction tends to increase with branch order along the vascular network (i.e. as the arteriole shows a smaller and smaller diameter) (Figures 16 and 17). In immediately pre-capillary arterioles of the bat wing, a relative loss of this relationship may reflect that these vessels become more highly sensitive to vasodilator metabolites. A quantitative measure of myogenic reactivity, *myogenic index (gain)* [164], provides a convenient tool for comparisons between vessels based on vessel diameter and not branch order. Shown in Figure 17, as arteriolar diameter decreases, there is an increase (in this case to more negative numbers) in the magnitude of the myogenic index [88].

5.2   SENSED AND CONTROLLED VARIABLES IN MYOGENIC CONSTRICTION

A difficulty with fully understanding the mechanistic basis for the myogenic response has been the lack of a clear knowledge of the sensed and controlled variable(s). As intraluminal pressure is increased in an arteriole, the circumferential orientation of the smooth muscle cells exerts an initial

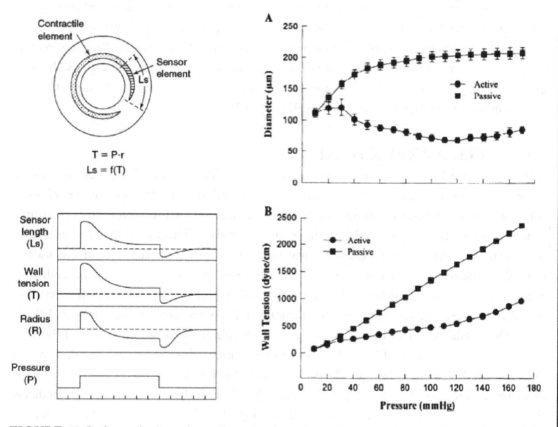

FIGURE 18: Left panels show the wall tension hypothesis for myogenic regulation as described by Johnson [205]. The right panels show pressure–diameter relationships (A) and wall tension–diameter relationships for isolated and cannulated rat cremaster muscle first order arterioles. Active and passive responses were determined in the absence on intraluminal flow [434].

passive stretch from their resting length (Figure 15). Despite this, the observation that some vessels constrict to below baseline diameter following an acute pressure step suggests that overt smooth muscle cell length is not the controlled variable. However, this does not discount the possibility that a membrane element remains stretched despite the pressure-induced contraction (as might occur if the sensory and contractile elements of the arteriolar VSMCs are arranged in series as described below). Further arguing against a requirement for overt cell stretch, arterioles subjected to a ramp pressure increase (largely negating the acute passive stretch observed to occur during a step increase in pressure) constrict to the same steady-state diameter as when an acute pressure step is applied [182].

On the basis that a length-based sensor did not fit with experimental observations, Johnson suggested that wall tension was the sensed variable. Using a model where a sensor element was placed in series with a contractile element (Figure 18), Johnson [205] demonstrated that steady-state diameter could be maintained at a level less than that prior to a applied pressure increase, if tension was the regulated variable. Tension was calculated according to the LaPlace relationship, tension = pressure × radius. Available in vivo and in vitro data support this concept as although myogenic constriction tends to normalize tension, its regulation is not complete leaving a sustained error stimulus even in the steady-state [47, 434] (Figure 18).

5.3   CELLULAR MECHANISMS

Although studied for over 100 years, the exact cellular mechanisms underlying pressure-induced vasomotor responses remain uncertain. In part, this relates to the parallel development of our knowledge with respect to vascular smooth muscle contraction, per se, and its modulation by extrinsic factors including the endothelial cells. Similarly, our knowledge of ion channel function, intracellular transduction mechanisms (for example, involving changes in intracellular Ca^{2+} and kinase activity) and the role of cytoskeletal and extracellular matrix elements have been continually developing. Armed with a new understanding of how these basic cellular functions work, it is now a challenge to understand how they are utilized in a mechanically induced vasomotor response (Figure 19).

The cellular basis for myogenic contraction lies within the smooth muscle cells of the arteriolar wall. Careful removal of the endothelium using a variety of methods (mechanical, air perfusion, controlled exposure to detergents) has clearly shown that myogenic constriction persists in the de-endothelialized vessel (Figure 20; [123, 236, 271]). The endothelium does, however, modulate the level of myogenic constriction. This occurs through the release of paracrine factors (for example, NO, prostaglandins and other metabolites of arachidonic acid) and direct electrical coupling via myoendothelial gap junctions (MEGJ) [340, 341, 363]. Activation of endothelial control mechanisms occur in response to various circulating and locally produced factors as well as some

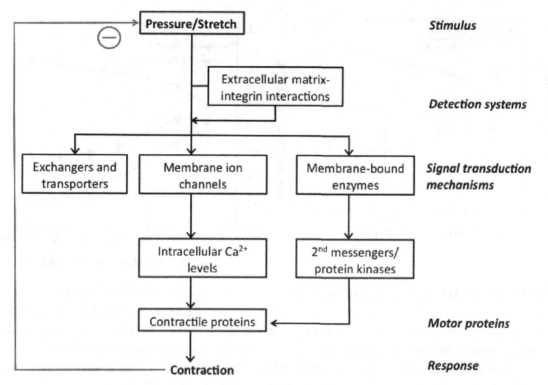

FIGURE 19: Schematic diagram illustrating the cellular events believed to contribute to myogenic vasoconstriction. A change in pressure exerts a force on arteriolar vascular smooth muscle cells—this may be direct or via interactions with the extracellular matrix. The mechanical stimulus is 'detected' by an as yet undefined sensor that then initiates a variety of signal transduction mechanisms. These biochemical reactions lead to activation of the contractile or motor proteins that cause contraction. In the case of a pressure increase, the contraction is thought to oppose the mechanical stimulus that provided the initial error signal.

mechanical stimuli (for example changes in shear stress) but does not appear to be directly affected by physiological levels of intraluminal pressure. This is illustrated in Figure 20 where pressure–diameter relationships were determined for single cannulated arterioles with and without endothelium (left panel) and in the presence and absence of a constant level of flow (right panel) [237]. In the latter case, flow results in a largely parallel shift in the pressure–diameter relationship to larger diameters (for each level of intraluminal pressure) consistent with the flow or shear-induced release of a vasodilator substance (in this case reported to be NO [237]).

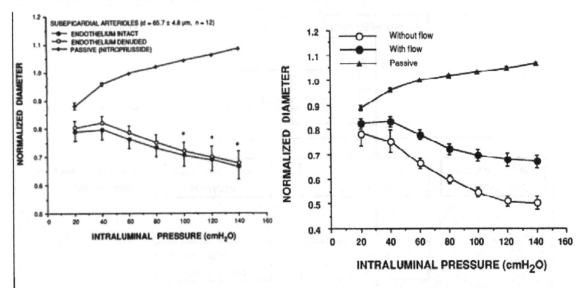

FIGURE 20: (A) Under isobaric conditions, mechanical removal of the endothelium does not alter the pressure–diameter relationship for cannulated porcine subepicardial arterioles. The data indicate that myogenic responsiveness is a function of the vascular smooth muscle cells. From Kuo et al. [237]. **(B)** A constant level of flow causes a parallel shift in the pressure–diameter relationship of cannulated porcine subepicardial arterioles. The data are consistent with the constant level of flow (shear stress) causing the release of a dilator compound (from the endothelium). From Kuo et al. [237].

To consider some of the important molecules underlying myogenic constriction, we will deviate from the likely temporal sequence of signaling events. The reader is referred to the general pathway shown in Figure 19 for context and reviews specific to the cellular signaling mechanisms underlying the arteriolar myogenic response [74, 87, 88, 176, 346, 363]. Additional information relating to candidate ion channels and signaling molecules is provided in Tables 3–5.

5.4 ROLE OF INTRACELLULAR Ca^{2+}

A key component in pressure-induced contraction, as for smooth muscle contraction in general, is the availability of a source to supply intracellular Ca^{2+}. Thus, early mechanistic studies demonstrated that arterioles placed in a physiological solution lacking Ca^{2+} (and often containing a divalent ion chelator such as EGTA) soon lost inherent myogenic tone and subsequently behaved passively in response to an increase in intraluminal pressure [106, 394, 395]. These observations also demonstrate that myogenic constriction requires an extracellular Ca^{2+} source and cannot be maintained,

CHANNEL	ION SELECTIVITY	INVOLVEMENT IN ARTERIOLAR MYOGENIC CONSTRICTION
Nonselective cation channels Stretch-activated channels (SACs)	$Na^+ < K^+ < Ca^{2+}$	SACs occur in vascular smooth muscle cells of arterioles and arteries. Opening of SACs at physiologically levels of membrane stretch results in a predominantly Na^+ current leading to membrane depolarization
TRPC1	Cations, non-selective	First proposed as a mechanosensor in oocyte expression systems. Role as a myogenic sensor unlikely given that TRPC1$^{-/-}$ mice have a normal phenotype and unaltered cerebral artery myogenic tone.
TRPC6	$Ca^{2+} > Na^+$ (5:1)	Oligonucleotides directed towards TRPC6 attenuates to pressure-induced cerebral artery SMC depolarization and myogenic constriction. Myogenic responsiveness and blood pressure are both, however, augmented in TRPC$^{-/-}$ mice – probably as a result of upregulation of TRPC3
TRPM4	Monovalent cations	Oligonucleotides directed towards TRPM4 (but not M5) inhibit pressure induced depolarization in cerebral artery smooth muscle. TRPM4 channels are selective for monovalent cations and modulated by both Ca^{2+} and PKC. Suggested to lie downstream of mechanical events activating other TRPs (e.g., TRPC6)
Na^+ channel ENac	Na^+	First identified as a mechanosensor in *C. elegans*. Later linked to shear-stress-mediated mechanotransduction in oocytes and blood vessels where ENaC might be important in the myogenic responses of rat cerebral arterioles and mouse renal interiobular arteries

TABLE 3: Putative plasma membrane ion channels mediating arteriolar myogenic vasoconstriction: their roles as primary 'mechanosensors'

Note: Re-printed from Hill et al. [180] with permission. See original source for details of contributing references.

		TABLE 3: *(continued)*		

CHANNEL	ION SELECTIVITY*	INVOLVEMENT IN ARTERIOLAR MYOGENIC CONSTRICTION
Ca^{2+} channels Voltage-Gated Ca^{2+} Channels	Ca^{2+}	Direct effects of stretch on VGCCs do not seem to be sufficient to explain the extent of Ca^{2+} entry. Ion channel phosphorylation subsequent to stretch activation modulates Ca^{2+} entry via VGCCs. Stretch induced depolarization persists in the presence of inhibitors of VGCCs
K^+ channels BK_{Ca}	K^+	Membrane stretch activates BK_{Ca} channels via an integrin-mediated mechanism. Opening of BK_{Ca} provides a hyperpolarizing current that opposes myogenic construction. Opening of BK_{Ca} might be regulated by pressure-stimulated generation of Ca^{2+} sparks. Closure of BK_{Ca} might result from pressure-induced production of 20-HETEs (Table 5)
K_v	K^+	Seem to have a negative feedback role in limiting myogenic reactivity. Inhibitors of K_v (4 aminopyridine, correolide) cause vasoconstriction of myogenically active small arteries
TREK-1	K^+	TREK-1 (TWIK-related K^+ channels) are activated by mechanical and chemical stimuli, including arachidonic and α linolenic acids. Operate at physiological levels of membrane potential. Basilar arteries from TREK-$1^{-/-}$ mice do not show responsiveness, suggesting that these channels might modulate vasoconstriction rather than determining levels of myogenic tone
Cl^- channels Cl_{Ca}	Cl^-	Involvement suggested on the basis of a favorable electrochemical gradient. Opening of Cl^- channels upon mechanical stimulation would lead to depolarization. Although largely pharmacological evidence supports this possibility, inhibitors with the necessary specificity are not currently available

TABLE 4: Putative 'mechanosensors' (non-ion channel in arteriolar myogenic vasconstriction)

CLASS	SPECIFIC ELEMENTS	INVOLVEMENT IN ARTERIOLAR MYOGENIC CONSTRICTION/ MECHANO TRANSDUCTION
Integrins	$\alpha_v\beta_3$, $\alpha_5\beta_1$	Myogenic constriction is impaired by disruption of smooth muscle cell interactions with ECM proteins, for example by RGB peptides or function-blocking antibodies. Evidence for ECM-integrin links modulating ion channel activity and Ca^{2+} handling. Evidence for an important role for fibronectin.
Cytoskeleton	Actin, microtubules	Myogenic constriction associated with G- to F-actin transitions. Depolymerization of microtubules causes vasoconstriction that seems to involve Rho-A-dependent Ca^{2+} sensitization without an overt increase in Ca^{2+}, Vimentin- and desmin-deficient mice show normal myogenic responses despite alteration in other properties.
G-proteins	$G\alpha_q$	Recent studies show agonist-independent activation of AT_1R via mechanical stimuli. Activation of $G\alpha_{2+}$ via this GPCR leads to generation of a second messengers (PLC, DAG, $Ins(1,4,5)P_2$) and stimulation of TRP proteins (including TRPC3, 6 and 7). Cation current leads to depolarization and contraction.
Caveolae	Caveolin 1	Myogenic constriction is impaired both by genetic deletion of caveolin-1 and acute chemical disruption of caveolae. Proposed mechanisms of action include alteration in Ca^{2+} handling, activation of BK_{Ca} and impaired activation of Rho through inhibition of translocation to caveolae.
Lipid bilayer	Intrinsic membrane proteins	Pressure-induced changes of the biophysical properties of the cell membrane might conceivably alter the mobility of intrinsic proteins and initiate signaling events.

Note: Re-printed from Hill et al. [180] with permission. See original source for details of contributing references.

TABLE 5: Intracellular signaling mechanisms in arteriolar myogenic vasoconstriction

MEDIATOR	TARGET(S)	INVOLVEMENT IN ARTERIOLAR MYOGENIC CONSTRICTION
Ca^{2+}		
Ca^{2+} entry	Calmodulin–MLCK	Activation of MLCK via calmodulin
Ca^{2+} release	R_yR, BK_{Ca}	Entry of extracellular Ca^{2+} activates SR ryanodine receptors and causes Ca^{2+} release. Increase in global Ca^{2+} required for force generation or also acts on local basis to regulate Ca^{2+} –sensitive ion channels
Lipid Mediators		
Diacylglycerol	PKC; TRPC3	Produced in a small renal arteries in response to increased intraluminal pressure. Activates some isoforms of PKC. Direct stimulatory effect on TRPC3
20-HETE	BK_{Ca}	Lipid mediator causing smooth muscle membrane depolarization by inhibition of BK_{Ca}
Ins $(1,4,5)P_x$	Ins $(1,4,5)$ P_3R	Produced in renal arteries in response to increased intraluminal pressure. Presumed to result from G-protein activation and acta on SR Ins $(1,4,5)$ P_3R receptors to release intracellular Ca^{2+}
Protein–kinase-mediated pathways		
MLCK	MLC_{20}	Activation of myosin facilitates acto-myosin cross-bridge cycling. Inhibition uncouples pressure-induced Ca^{2+} increases from contraction.
PKC	Multiple intracellular targets including ion channels, MAP kinase and CPI-17	Modulation of Ca^{2+} sensitization through activation of CPI-17. Current evidence largely indirect (relying on pharmacological inhibitors. Also activates membrane channels (e.g., VGCC) and other kinases (e.g., p24/44 MAP kinase)

Rho kinase	Includes MYPT1	Ca^{2+} sensitization via phosphorylation of MYPT1 and inhibition of myosin phosphatase. Potentiates contraction via slowing and/or preventing diphosphorylation of P-MLC$_{20}$
Sphingosine kinase	Sphingosine	Membrane sphingomyelin-derived sphingosine is phosphorylated to S-1P to act as a second messenger acting via receptors coupled to G proteins, phospholipase C and Rho kinase. Implicated in myogenic constriction given that S-1P is activated by depolarization and subsequently stimulates both SR Ca^{2+} release and Rho-A–mediated Ca^{2+} sensitization.
Tyrosine phosphorylation	Multiple intracellular targets including ion channels, MAP kinases	A variety of mechanical stimuli initiate protein trycosine phosphorylation including cSRC and p-42/44 MAP kinases. These events modulate ion channels, transduce integrin-mediated events, reorganize cytoskeletal proteins and mediate remodeling of the vascular wall. Current literature favors a modulatory or facilitatory role in myogenic constriction rather than being obligatory
Myosin phosphorylation	P-MLC$_{20}$	Desphosphorylation of P-MLC$_{20}$. Inhibition of contraction. Target of Rho kinase and CPI-17
Reactive oxygen species		
H_2O_2	BK$_{Ca}$ to be defined	Increased expression of NADPH oxidase components in arterioles relative to arteries. Myogenic reactivity of rat tail arterioles inhibited by catalase and in mice with targeted impairment of p47(phox) or Pac1. By contrast, H_2O_2 acts as a vasodilator in many vascular preparations via activation of a smooth muscle BK$_{Ca}$, suggesting that it is unlikely to be a general mediator of myogenic vasoconstriction

Note: Re-printed from Hill et al. [180] with permission. See original source for details of contributing references.

alone, by a cyclical release of Ca^{2+} from intracellular stores or by mechanisms not requiring intracellular Ca^{2+}.

The pathway underlying pressure-induced Ca^{2+} entry (Figure 19) is currently thought to involve an initial membrane depolarization mediated via a class, or classes, of non-selective cation channels which leads to an increased opening probability of the L-type voltage-gated Ca^{2+} channel [221, 232]. That the depolarization occurs upstream of opening of the voltage-gated Ca^{2+} channels has been directly demonstrated by studies measuring pressure-induced changes in membrane potential (Em) in the presence of selective Ca^{2+} channel blockers (including verapamil, nifedipine and nisoldipine) [221, 232]. Although the L-type voltage-gated Ca^{2+} channel has been shown to have some inherent mechanosensitivity, and can be activated by extracellular matrix protein–integrin interactions [422], it appears that any direct effect of pressure on the channel plays only a modulatory role [269].

Measurements of pressure-induced changes in SMC global Ca^{2+}_i (using the ratiometric fluorescent indicator Fura 2) suggest that over the myogenic pressure range (for example, 30–120 mm Hg in a cannulated skeletal muscle arteriole) Ca^{2+} levels vary from approximately 100–250 nM [434]. This contrasts to Ca^{2+}_i levels of approximately 60 nM in the passive state [434]. Similar changes in Ca^{2+}_i were also shown in small cerebral arteries under in vitro and cannulated conditions [221]. Several studies have argued [1] that as the active state pressure-induced changes in Ca^{2+} are comparatively small (relative to that induced by an agonist) and [2] that the greatest change in Ca^{2+} is seen as vessels transition from a passive to active state at low pressures, mechanisms other than a simple rise in Ca^{2+}_i level must be involved [308]. In particular, it has been proposed that changes in the sensitivity to Ca^{2+}_i occur such that a greater contractile activation occurs for a given change in Ca^{2+} [308] (see also, *Myosin Light Chain Phosphorylation—Dual Regulation by Myosin Light Chain Kinase and Myosin Phosphatase*).

5.5 INVOLVEMENT OF INTRACELLULAR CA^{2+} RELEASE AND OTHER CA^{2+} ENTRY MECHANISMS?

In addition to Ca^{2+} entry from the extracellular space, Ca^{2+} can be mobilized from intracellular stores to contribute to a stimulus-induced intracellular Ca^{2+} signal. While Ca^{2+} can be stored in a number of intracellular compartments (including the SR, mitochondria, nucleus and bound to Ca^{2+} binding proteins), the most relevant to myogenic constriction is the SR pool. This is based on the SR's well-described role in agonist-induced SMC contraction and the kinetics of Ca^{2+} release from, and re-uptake into, this compartment. Further, studies have shown in isolated and cannulated arteries that [1] increased intraluminal pressure leads to an accumulation of inositol phosphates and diacylglycerol [296] and [2] inhibition of phospholipase C inhibits myogenic contraction [310]. While indirect, these studies implicate a role for intraluminal pressure-induced production of IP_3 and therefore release of Ca^{2+} from the SR by activation of IP_3R. However, as myogenic constriction still occurs in some vascular beds after depletion of SR Ca^{2+} stores (using ryanodine or SERCA pump inhibitors), there is doubt whether store release is obligatory. Isolating the direct role or SR release in an acute myogenic contraction is made difficult by the fact that removal of Ca^{2+} from the extracellular compartment both causes a loss of myogenic tone and often depletes intracellular stores (these effects are further exacerbated when a chelator such as EGTA is used to remove trace amounts of extracellular Ca^{2+}).

SR Ca^{2+} release may, however, modulate myogenic reactivity through its involvement in spatio-temporal components of Ca^{2+} signaling. For example, the SR underlies the generation of both Ca^{2+} sparks and Ca^{2+} waves. Ca^{2+} sparks typically arise from spontaneous release events through ryanodine-sensitive SR membrane channels [299] while Ca^{2+} waves are dependent on cyclical release of Ca^{2+} through IP_3-sensitive channels. The exact role of these events in myogenic signaling, however, remains uncertain as Ca^{2+} sparks have been reported to oppose pressure-induced contraction (through activation of BK_{Ca}) [299], facilitate myogenic contraction (through increasing in frequency and summing to intracellular Ca^{2+} waves) [391] and be absent in some vessels despite exhibiting active myogenic tone [416, 426]. Similarly, while Ca^{2+} waves increase in frequency in response to acute increases in intraluminal pressure, they persist when myogenic tone is inhibited and their absence does not prevent the development of arteriolar myogenic tone [117]. Although further studies are required to determine whether the above differences reflect vascular heterogeneity or methodological differences, it is unlikely that these events are fundamental to arteriolar myogenic signaling, although they may play important modulatory roles. Alternatively, some of the second messengers may serve novel roles such as the recently demonstrated ability of activated IP_3 receptors to directly modulate plasma membrane Trp and BK_{Ca} channels independently of SR Ca^{2+} release [4, 5, 431].

A number of other mechanisms of Ca^{2+} entry could conceivably contribute to the development of myogenic tone. Included among these are Ca^{2+} entry via the reverse-mode action of the sodium–calcium exchanger (NCX1) [322, 430]; through T-type voltage-gated channels (including novel splice variants) [235]; or Ca^{2+} permeable non-selective cation channels. Clearly, such mechanisms do not solely account for the Ca^{2+} required to activate the contractile proteins, as selective inhibitors of L-type voltage-gated channels are generally effective in negating myogenic tone in most arterioles and small arteries that have been studied [179, 222, 232]. Ca^{2+} entry via these other mechanisms may, however, indirectly affect myogenic tone through, for example, local regulation of Ca^{2+}-sensitive ion channels.

5.6   MEMBRANE POTENTIAL

A relationship between arteriolar intraluminal pressure and membrane potential has been shown, using sharp glass microelectrodes, in several isolated arteriole preparations including those from the

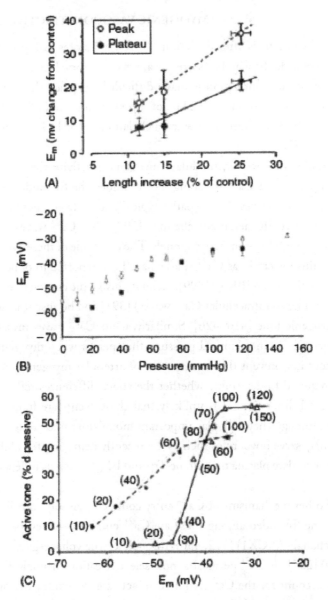

FIGURE 21: Relationships between smooth muscle cell membrane Em and extent of mechanical stimulation (stretch (A); pressure (B and C)). Panel A illustrates the change in Em in single smooth muscle cells stretched along their longitudinal axis. Data are plotted for the peak response and at steady-state. From Reference [86]. Panel B compares the relationship between intraluminal pressure and VSM cell Em for cannulated rat cerebral (open symbols) and cremaster muscle (closed symbols) arteries. Em was determined using glass microelectrodes. From Reference [232]. Panel C shows the data from Panel B plotted relative to the degree of active tone developed at a given intraluminal pressure (pressures are shown in parentheses). These data demonstrate a sigmoidal relationship between Em and the extent of myogenic contraction. Further this relationship is shown to differ between tissues.

cerebral and skeletal muscle circulations (Figure 21) [165, 221, 232]. Similar results have also been shown using glass electrode recording techniques in vivo preparations [420].

Collectively, these studies demonstrate a number of interesting and important points in regard to the relationship between SMC membrane potential and the level of myogenic tone. Firstly, at physiologically relevant levels of intraluminal pressure, arteriolar smooth muscle is depolarized to approximately −40 to −30 mV. This level of depolarization is consistent with a membrane potential which favors the opening of L-type VGCCs. Secondly, after a relatively linear increase of Em with

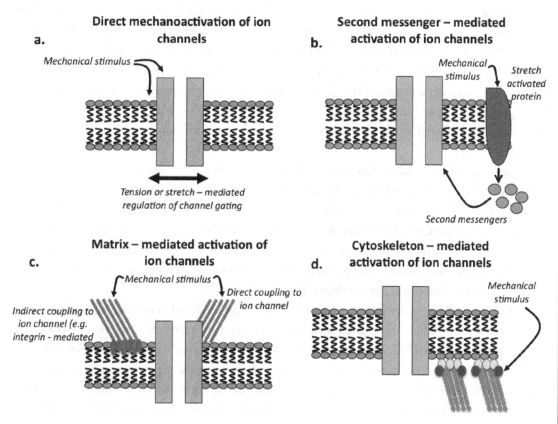

FIGURE 22: Possible mechanisms by which stretch or tension activates ion channel signaling in arteriolar smooth muscle cells. Panel a. Mechanical stimulus acts directly on ion channel subunits to affect an alteration in channel gating. Panel b. Mechanical stimulus acts on a membrane (e.g., receptor) or a membrane-associated protein (e.g., a trimeric G-protein) to generate second messengers (IP$_3$, DAG, Ca^{2+}) which subsequently modulate ion channel gating. Panel C. Mechanical stimulus deforms ECM proteins which in turn affect ion channel gating directly or secondarily to binding proteins such as integrins. Panel d. Mechanical stimulus affects alterations in the cytoskeleton which in turn leads to the activation of membrane ion channels. Modified from Hill and Meininger [178].

intraluminal pressure, the relationship tends to plateau such that as pressure continues to increase, only a relatively small change in Em is observed. This latter observation has been used as additional, although indirect, evidence to support the concept that myogenic constriction may involve mechanisms other than depolarization-induced Ca^{2+} entry via VOCCs (Figure 22) [308].

5.7   MECHANISMS CONTRIBUTING TO PRESSURE-INDUCED MEMBRANE DEPOLARIZATION

While the exact nature of the non-selective cation channel leading to pressure-induced membrane depolarization is uncertain, several candidates have been proposed. Considerable interest has been placed on the possible role of a plasma membrane stretch-activated channel (SAC). Using patch clamp techniques in combination with either direct cell stretch or deformation of the membrane via suction (applied via the patch pipette) or hypotonic buffer-induced swelling, mechanically activated currents have been demonstrated [86, 351, 415]. Importantly, in studies directly stretching VSMC, the degree of stretch required to activate the channels was consistent with the extent of cell stretch that would be predicted in the intact vessel when intraluminal pressure was altered across physiological levels [86].

A key question remains as to how is a change in intraluminal pressure linked to the opening of NSCC. Is there a direct link to membrane deformation as might be predicted from the existence of SACs? In this case, the channel proteins may be inherently mechanosensitive or their local environment is mechanically changed leading to an alteration in their conformation and gating properties. Does the linkage require tethering to extracellular (for example to extracellular matrix proteins) or cytoskeletal elements? Alternatively, is activation indirect requiring the generation of intracellular signals, or second messengers, which then activate the NSCC?

Recent studies consistent with the second messenger-induced activation of NSCC can be divided into two possibly inter-related hypotheses. The first considers the activation of Trp channels via Ca^{2+}, DAG or PKC while the second hypothesis suggests that the mechanosensitivity of GPCRs leads to the activation of phospholipases and the subsequent production of second messengers that activate NSCC [314, 364]. In both situations, the mechanically activated second messenger pathways lead to membrane depolarization and opening of voltage-gated Ca^{2+} channels.

Early evidence supporting the involvement of second messenger pathways in myogenic constriction was provided by pharmacological inhibitor studies implicating roles for G-proteins, phospholipase C and protein kinase C [177, 309, 310]. These were supported by direct measurements of intraluminal pressure-induced accumulation of DAG and inositol phosphates in small, cannulated, renal arteries [296]. More recently, evidence has been provided for mechanosensitivity of G protein-coupled receptors (GPCRs) including the angiotensin II receptor [273]. A direct mechanical effect of pressure on the GPCRs was demonstrated to be agonist independent although in the example of the AT1 receptor, pressure-induced activation of $G\alpha_{q/11}$ could be inhibited by losartan a specific AT1 receptor blocker. It was proposed that mechanical force directly alters the conformation of the receptor proteins inducing an activate state. Interestingly, this mode of mechanical activation is not limited to the AT1 receptor as it could also be demonstrated for several other GPCRs including those for histamine and vasopressin. The effect was, however, not totally non-selective as mechanical activation of the Gs-coupled β_2 adrenoceptor could not be demonstrated.

As mentioned above, subsequent activation of phospholipases and generation of second messengers may then lead to modulation of ion channels and levels of membrane potential. Consistent with this sequence of events, phospholipase-based signaling has been implicated in the activation of cation channels including those comprised of Trp proteins. For example, activation of TrpC6 currents could conceivably lead to membrane depolarization, opening of VGCC and myogenic contraction. This would, therefore link a number of earlier observations, including that acute TrpC6 knockdown inhibits myogenic responsiveness [414] without the Trp channel having to be mechanosensitive per se. Similarly, TrpM4 has been implicated in myogenic signaling and is activated by events downstream of phospholipase activation [113]. As these events can also occur subsequent to classical agonist-induced signaling, conceivably, they may interact with myogenic signaling via stretch-induced release of agonists [213] or synergistic intracellular signaling mechanisms [251]. Additional information on trimeric G-protein involvement in myogenic signaling can be found in recent reviews by Storch et al. and Kauffenstein et al. [213, 373].

An alternative mechanosensory pathway involves force transmission via extracellular matrix proteins (ECM). Thus, ECM protein binding to cell membrane integrins provides a direct link from the extracellular matrix environment and a mechanism for bi-directional transmission of physical forces. Additionally, integrins can act as recipients of specific matrix protein-coded information and assist in signal amplification (outside-in and inside-out signaling). As an example, physical forces may impact the outside-in pathway via strain applied to existing linkages or may cause conformational changes in (and unfolding of) ECM proteins to uncover cell membrane binding motifs. The latter concept is supported by studies demonstrating 'matricryptic' sites buried in the three dimensional structure of native matrix proteins [81] and those showing that specific matrix proteins such as fibronectin possess domains that can be extended on application of force [119].

Specifically implicating a role for integrins in arterioles, integrin-recognizing synthetic RGD peptides decrease VSMC intracellular Ca^{2+} and cause vasodilation [77, 284]. Demonstrating specificity, these actions are prevented by function blocking antibodies directed at the β_3 integrin subunit [284]. Further support for integrins as sensors involved in the myogenic response was similarly provided by studies of isolated arterioles showing that blockade of $\alpha_v\beta_3$ or $\alpha_5\beta_1$ integrins, with antibodies, abolishes myogenic constriction to step increases in intraluminal pressure [264]. Other important evidence for integrins being possible initiating events in arteriolar mechanotransduction is that integrin activation leads to phosphorylation and activation of ion channels (including VGCCs and BK_{Ca}) previously demonstrated to be important to myogenic responsiveness [421, 423].

Involvement of matrix proteins in myogenic signaling is likely more complex than the outside-in directed signaling through integrins described above. Matrix proteins are known, for example, to bind cell surface proteins other than integrins. In addition, integrin binding affinity can be modulated by intracellular mechanisms [217], raising the possibility that integrin–ECM protein interactions may be modulated by inside-out signaling mechanisms during myogenic or agonist-induced changes in activation. In this case, integrin involvement may be viewed as an adaptive response to the contraction, perhaps reinforcing or stabilizing the contractile state and adhesion sites.

5.8 MYOSIN LIGHT CHAIN PHOSPHORYLATION—DUAL REGULATION BY MYOSIN LIGHT CHAIN KINASE AND MYOSIN PHOSPHATASE

Following pressure-induced increases in intracellular Ca^{2+}, contraction occurs via Ca^{2+}/calmodulin-dependent activation of myosin light chain kinase and phosphorylation of the regulatory 20 kDa myosin light chain subunit. Phosphorylation of ser 19 of the regulatory light chain causes a conformational change in the myosin head group region that facilitates its interaction with actin such that ATP-dependent crossbridge cycling can occur. Evidence for the importance of myosin phosphorylation in arteriolar myogenic constriction has been provided by studies using myosin light chain kinase inhibitors (for example, ML7; [434]) and direct measurements of light chain phosphorylation using both 2D electrophoretic [434, 435] and Western blotting [116, 211] approaches.

Using cannulated rat cremaster muscle arterioles, measurements of MLC phosphorylation (using 2D electectrophoresis) in response to an acute pressure step showed a rapid increase in phosphorylation (within 10 sec) that was maintained despite attainment of a steady-state level of constriction [434, 435]. Further, the level of MLC phosphorylation was shown to be directly related to the calculated (according to the Law of LaPlace) level of wall tension. Direct measurements of pressure-induced myosin light chain phosphorylation were subsequently demonstrated in both cerebral small arteries [116, 211] and cremaster muscle arterioles [287] using a more sensitive Western blotting approach.

In addition to pathways mediating an overt increase in intracellular Ca^{2+} levels, mechanical events likely also activate mechanisms that increase the Ca^{2+} sensitivity of the contractile process. A particular mechanism that mediates such an action is via inhibition of myosin phosphatase. Inhibition of the phosphatase would be predicted to slow the breakdown of the phosphorylated form of MLC and thus enhance contractile activity (see Figures 6 and 7). To date, two mechanisms have been demonstrated, in vascular smooth muscle, to underlie inhibition of the phosphatase, Rho kinase-mediated phosphorylation of the myosin phosphatase targeting subunit (MYPT1) and PKC-mediated activation of CPI-17 [74].

Initial studies implicating a role for RhoA/Rho kinase-mediated Ca^{2+} sensitization and the myogenic response examined the effect of pharmacological inhibitors (Y-27632, HA-1077 and H1152) on pressure-induced arteriolar tone. In rat pressurized mesenteric vessels, Y-27632 abolished myogenic tone without decreasing Ca^{2+}_i [398]. Schubert et al., using rat-tail resistance arteries, reported similar concentrations of the inhibitor attenuated, but did not abolish, myogenic tone [345] although the inhibitory effect was again not associated with a decrease in global Ca^{2+} levels. Using the hydronephrotic rat kidney model, Nakamura et al. showed both Y-27632 and HA-1077 to dilate afferent arterioles while having little effect on efferent arterioles [294] which are known to have little myogenic responsiveness [115, 257, 354].

More recently, improvements in the sensitivity of Western blotting techniques have enabled direct measurements of MYPT1 phosphorylation in arterioles [116, 211, 381]. These authors have shown that MYPT1 phosphorylation occurs at threonine 855 in myogenically active cerebral arteries and that this contributes to pressure-induced MLC phosphorylation [211]. Further, MYPT1 phosphorylation was shown to mediate the interaction that occurs between intraluminal pressure and an applied agonist (specifically, serotonin) [116]. Importantly, these studies provide direct biochemical measurements for mechanisms underlying Ca^{2+}-sensitization in pressurized resistance vessels.

As with the activation of NSCCs, the link between an increase in intraluminal pressure and activation of Rho kinase currently remains unclear. Evidence has been provided for a possible role for pressure-induced production of sphingosine 1-phosphate (S1P) acting through the S1P receptor [36]. Alternatively, Rho kinase is known to be activated downstream of focal adhesion formation possibly implicating a role for ECM and cytoskeleton-related events. The latter is an attractive proposition as it combines several elements that have individually been implicated in myogenic signaling.

Contributions from a number of other signaling molecules and pathways to myogenic constriction have been identified [87, 88, 104, 116, 211, 213, 346, 381]. Whether this diversity of mechanisms reflects regional or species differences, or which act through pathways that converge on common elements (for example, altering properties of the plasma membrane, cytoskeleton, contractile protein interactions, or Ca^{2+} handling) remains to be clarified. Importantly, as well as for specific roles in mechanotransduction, a number of these mechanisms are currently being studied in regard to their involvement in smooth muscle contraction, per se.

5.9   NEGATIVE FEEDBACK MECHANISMS REGULATING MYOGENIC CONSTRICTION

Considerable argument was historically made as to whether myogenic constriction could act in a feed-forward manner leading to vascular instability and contributing to increased pressure upstream of the myogenically active vascular segment [163, 223]. An argument against this was that arterioles exhibit myogenic reactivity across a limited pressure range; passively collapsing when distending pressure is too low and showing forced dilation beyond the myogenic pressure range. Further, not all arterioles and small arteries within a given vascular network exhibit the same degree of myogenic reactivity. While smaller arterioles have been shown to exhibit stronger myogenic responses, the degree of myogenic responsiveness tends to decline in the more proximal vessels of the network [84]. In addition to these physical limitations, evidence has more recently been provided for second messenger-based activation of negative feedback mechanisms.

Nelson and colleagues demonstrated the presence of intracellular Ca^{2+} sparks in smooth muscle cells of small cerebral arteries [299]; this subcellular Ca^{2+} transient having been previously identified in cardiac muscle [61]. The Ca^{2+} sparks resulted from focal SR release via ryanodine receptors [299]. Further, the resultant Ca^{2+} release occurred in a restricted space between the SR and plasma membranes raising local (sub-membranous) Ca^{2+} levels to concentrations nearing 10 μM. This increase in Ca^{2+} is within the range required to influence to the opening of Ca^{2+}-activated ion channels, in particular, the large conductance Ca^{2+}-activated K^+ channel (BK_{Ca}). Thus, an attractive hypothesis was that pressure-induced mobilization of Ca^{2+}, in addition to activating contraction, stimulates ryanodine receptor-mediated spark production thereby activating BK_{Ca} and a resultant outward current, hyperpolarizing the membrane and opposing VOCCs.

A caveat with this hypothesis is whether it is applicable to all vascular beds. While the evidence for Ca^{2+} sparks, linked to BK_{Ca} activation, is strong in cerebral VSMCs, recent studies have cast doubt on their role in smooth muscle cells from arterioles of skeletal muscle [416, 426]. Skeletal muscle arteriolar smooth muscle appears to lack Ca^{2+} sparks. Further, difference appears to exist at the level of the level of the BK_{Ca} channel such that in VSMC from skeletal muscle it is less sensitive to Ca^{2+} and demonstrates a decreased probability of opening as compared to cerebral VSMCs [425, 426]. These differences collectively suggest possible heterogeneity in the regulation of this hyperpolarizing feedback mechanism. Further, they illustrate that similar signaling mechanisms may be regulated somewhat differently to obtain appropriate local responses.

5.10 DOES MYOGENIC VASODILATION SIMPLY REFLECT THE REVERSAL OF MYOGENIC CONSTRICTION?

Few studies have directly considered mechanisms underlying myogenic vasodilation that occurs as intraluminal pressure is reduced. Important questions remain including, what are the relative roles of accelerated uptake or removal of intracellular Ca^{2+} during myogenic vasodilation? Similarly, do Ca^{2+} de-sensitization mechanisms contribute? It is generally accepted that a decrease in pressure leads to a reduced mechanical stimulus for Ca^{2+} mobilization and that Ca^{2+} is removed by both sequestration and extrusion mechanisms. To date, the possibility that a decrease in pressure activates Ca^{2+} removal systems, for example, through the stimulation of cyclases and generation of cyclic nucleotides and/or modulation of SR Ca^{2+} uptake by molecules such as phospholamban, has not been extensively studied. Wellman et al. [413], using cerebral vessels from a phospholamban knockout mouse, showed that Ca^{2+} sparks and subsequent BK_{Ca} channel activity are increased by a PKA-dependent mechanism involving phospholamban. Further evidence that such mechanisms exist in arterioles is provided by studies showing that vasodilatory stimuli, working through increases in cyclic guanosine monophosphate (cGMP), have been reported to decrease myogenic tone through a mechanism involving a decrease in smooth muscle Ca^{2+} sensitivity [397]. An additional consideration is that if myogenic vasoconstriction is dependent on multiple temporally dependent pathways, then the rate of relaxation may differ from that of constriction due to differences in the time necessary to reverse particular processes. For example, if after a rapid onset MLC-dependent constriction, the steady-state mechanical response is maintained through Ca^{2+} sensitization, or through alterations in the assembly of the cytoskeleton, relaxation may vary markedly with respect to the initial rate of constriction. This may be compounded further if prolonged myogenic constriction were to lead to early remodeling events within the vascular wall [266]. While direct evidence does not exist for mechanistic differences in myogenic constriction and relaxation, it does, however, appear that changes in arteriolar smooth muscle Ca^{2+}_i and diameter following an acute decrease in intraluminal pressure are less consistent than those following a pressure increase [434], perhaps suggesting that this phase of myogenic responsiveness deserves consideration in its own right.

· · · ·

CHAPTER 6

Metabolic Control of Blood Flow

The metabolic hypothesis for the local regulation of blood flow suggests that a factor, or more likely factors, generated as a result of tissue metabolism cause(s) dilation of pre-capillary arterioles, increasing blood flow with subsequent convective washout/removal of the metabolite(s). The increase in blood flow further supplies additional nutrients, including oxygen, to meet the demands of an increase in metabolism. Vasodilator factors that have been implicated in metabolic control are numerous and include adenosine, K^+, lactate, ATP, CO_2, prostaglandins, hyperosmolality, and NO/EDHF [88].

Metabolic control may not be limited to situations where there is an increase in tissue activity (for example as occurs during muscle contraction). Metabolic regulation conceivably also contributes to blood flow autoregulation occurring during changes in perfusion pressure. Thus, as pressure increases (in the face of unchanged metabolic demand) the resultant changes in blood flow would lead to a relative increase in the provision of nutrients and removal of metabolites. Similarly, as perfusion pressure decreases, there would be a decrease in nutrient supply and a decrease in washout of metabolites. In terms of the magnitude of metabolite accumulation it has, however, been predicted that accumulation would be much greater during functional hyperemia (i.e., during increased tissue activity) than blood flow autoregulation [241].

Although it is likely that no single factor can fully account for metabolic vasodilation (and that differing factors may contribute to differing temporal components of the response—for example, initiation vs maintenance), in the following sections we will briefly consider evidence for some of the individual factors that have been proposed as mediators of metabolic vasomotor control. It should, however, be realized that metabolic factors may differ between tissues and may differ depending on the nature and magnitude of the stimulus. For example, similar acute alterations in perfusion pressure can occur under widely varying tissue oxygen tensions thus leading to potentially different contributions of the metabolic component to autoregulation of flow.

In general, to be considered a putative metabolic mediator (of vasodilation), the following criteria have been proposed [69]:

1. The factor (or a metabolite of the factor) should be produced in the parenchymal tissue and released to act on arterioles/resistance vessels.

2. The interstitial concentration of the factor should increase in proportion with tissue activity (this should be distinguished from a level of production that occurs both basally and during increased tissue activity).

3. Increased interstitial concentrations of the factor should precede the vasodilation.

4. Exogenous application of the factor should cause rapid vasodilation.

5. Inhibiting the production of the substance, or its action on the vasculature should attenuate the expected increase in blood flow.

A number of candidate metabolic factors are briefly described in the following sections. For additional information on these and other putative metabolic vasodilators in specific tissues, the reader is referred to several reviews [88, 96, 204, 392].

6.1 pO2

It may seem intuitive that alterations in levels of tissue oxygen (pO2) would be an appropriate error stimulus feeding back to control vascular resistance, blood flow and hence oxygen supply. However, oxygen supply at the tissue level is complex, involving the degree of arterial saturation, hemoglobin concentration, organ blood flow, arterial-venous shunting and capillary surface area. Thus, an absolute level of pO2 needs to be considered against these variables. Consistent with a role for pO2 in control of local blood flow is that in the systemic circulation hypoxia causes vasodilation while hyperoxia leads to vasoconstriction.

In regard to a significant role for pO2 levels in autoregulation following physiological changes in perfusion pressure, it appears unlikely that tissue levels of pO2 change sufficiently to elicit a vasomotor response. The situation is, however, likely to be different when there are concurrent changes in tissue activity and pO2 utilization. Supporting this concept, several whole organ studies (in skeletal muscle, coronary and cerebral circulations) have demonstrated an enhanced autoregulatory capacity when pO2 is maintained at low levels [212, 231]. Interestingly, and in contrast, increasing oxygen levels via a hyperbaric chamber did not impair autoregulation, leading those investigators to cast doubt on a primary role for pO2 [37].

If tissue pO2 is to be a determinant of autoregulatory responses, key questions are how are declining levels of O_2 detected and are their effects direct or indirect? Further, where are declining pO_2 levels sensed—within the tissues or in the vessel wall, itself? If in the vessel wall, what are the relative roles of the smooth muscle and endothelial cells? Direct effects could occur through the O_2 sensitivity of various molecules involved in smooth muscle cell contraction—ion channels (although considerable heterogeneity has been observed in regard to direct effects) [73, 138], regulatory enzymes and contractile proteins as well as enzymes related to metabolism [419]. Even in the case of a number of these, the effect could be indirect such as K_{ATP} channels being regulated by

ADP/ATP ratios [321] or via products of arachidonic acid metabolism resulting from the O_2 sensitivity of enzymes such as cyclooxygenase, lipoxygenase, and cytochrome P-450 monooxygenases [166]. Importantly, these enzymes exhibit a requirement for higher pO_2 levels than do the mitochondrial metabolic enzymes, suggesting that the former may be stimulated by pO_2 levels without having to achieve a reduction to levels critical for cellular metabolism. A final consideration is that there appears to be marked regional and species differences in the mechanisms through which tissue pO_2 is coupled to local blood flow control, again perhaps casting doubt on the role of O_2 as the critical metabolic factor.

6.2   ADENOSINE

At the cellular level, several adenosine receptors have been identified (A1, A_{2A} and A_{2B}) with its vasodilator effect being primarily exerted through the A_2 receptors in coronary [173], skeletal muscle [323] and cerebral [240] vasculatures. Adenosine has been shown to cause vasodilation both by a direct action on vascular smooth muscle and via endothelium-dependent mechanisms (via NO production). Differences in receptor-mediated pathways may reflect differences in stimuli (for example hypoxia as compared to hypotension), vascular beds and species studied.

Adenosine was initially shown to be a powerful coronary vasodilator that is released during hypoxia [27] and during hyperemic responses [335]. It was proposed that increased myocardial oxygen consumption was associated with adenosine release from cardiomyocytes. Subsequently, Rubio and Berne [334] showed that adenosine was released into interstitial fluid and suggested that its concentration regulated coronary blood flow to provide appropriate levels of tissue oxygenation. The resulting increase in blood flow would then restore tissue oxygenation, decreasing the further release of adenosine. Using chronically instrumented dogs, Tune and colleagues have evaluated the effects of exercise on coronary venous adenosine levels and on adenosine receptor blockade. Venous adenosine levels neither reached levels considered to be necessary for effective vasodilation nor overcame the effects of adenosine receptor blockade [393]. These authors have further argued that adenosine receptor blockade during exercise shows inhibition of a tonic vasodilator component rather than a synergistic effect between myocardial oxygen consumption and adenosine production. In contrast, adenosine does appear to play an important role during periods of myocardial ischemia [245] or during severe myocardial hypoperfusion [109]. Similarly, in studies of human skeletal muscle blood flow, adenosine appears to play more of a role in hypoperfused states [52, 53]. Thus, at this point in time, the strongest support for adenosine as a metabolic vasodilator comes from preparations under relative hypoxia or at the lower ends of the autoregulatory pressure range.

On the basis of the available evidence, it appears that adenosine does not meet the criteria for being the metabolite underlying autoregulation, despite (1) increasing rapidly with tissue activity,

(2) exerting vasodilator actions when exogenously administered, and (3) showing a time course of production consistent with changes in vascular resistance.

6.3   ATP

In addition to adenosine, adenine nucleotides have been suggested to be metabolic vasodilators. Consistent with this, ATP is released from contracting skeletal muscle, particularly during hypoxia, and is increased in venous blood during exercise. ATP is a known vasodilator, acting on endothelial cell P_2y receptors to increase $[Ca^{2+}]_i$ and cause the release of paracrine factors including NO, prostaglandins and EDHF [23, 120]. It should be noted that ATP applied directly to smooth muscle causes contraction via P_{2x} receptors [121]. Further support for a role for ATP comes from the observation that brachial artery infusion of exogenous ATP in humans causes a vasodilation of similar magnitude to that seen during maximal forearm exercise [105]. Although, in humans subjects, there is considerable uncertainty as to the actual mediator that underlies ATP-induced dilation [76].

Recent interest has arisen in the possibility that, in addition to release from skeletal muscle, red blood cells (RBCs) release ATP during periods of increased demand [365]. Supporting this, hypoxia (in the presence of hypercapnia) was shown to cause a rapid release of ATP from endothelial cells [24]; an observation that was later extended to more physiological situations [118]. Also of relevance to the microcirculation is that RBCs release ATP when undergoing deformation [366], as is required during transit through capillaries. Similar to the above, ATP would act on adjacent endothelial cell P_2Y receptors to cause production of paracrine vasodilator substances and subsequent vasodilation. It has been argued that such a scenario would allow coupling of red cell distribution (and hence O_2 carrying capacity) to metabolic requirements [365]. Perhaps arguing against RBC ATP as the principal metabolic stimulus in local blood flow regulation, (1) autoregulatory behavior can be seen in preparations in the absence of blood perfusion and (2) exercise hyperemia persists in subjects with cystic fibrosis where RBC release of ATP has been shown to be impaired [343].

6.4 K^+

As early as 1941 Dawes concluded that in the perfused hindlimb 'small doses of KCl cause vasodilation, larger doses vasoconstriction'. Further, he suggested 'the release of potassium ions is the cause of the vaso-dilation which occurs in contracting muscle' [92].

During muscular contraction K^+ is released from active fibers, via the opening of voltage-dependent K+ channels during repolarization, raising interstitial K^+ levels to levels between 5 and 20 mM [185]. At these concentrations extracellular K^+ acts as a vasodilator due to its activation of the inward rectifier K^+ channel, Kir, subsequent hyperpolarization and vascular smooth muscle relaxation. Note that at higher $[K^+]$ (approximately >25 mM K^+ will cause depolarization, opening of VGCC and contraction. Support for K^+ as a metabolic vasodilator has been strongest in studies of skeletal muscle where K^+ is released during very brief bouts of exercise (in the order of 1 second duration), during which vasodilation can also be demonstrated [10, 70, 185]. Further, inhibiting K^+ release attenuates this dilation [10].

In studies of isolated cerebral arterioles, McCarron and Halpern [270] showed that K^+ (5 mM added to a K^+ free superfusate) caused a transient dilation while the dilation was sustained at K^+ concentrations of 7–15 mM. The transient dilation was shown to be inhibited by oubain, while the sustained phase was inhibited by low concentrations of Ba^{2+}. Oubain is an inhibitor of the Na^+/K^+-ATPase while at the concentration used Ba^{2+} is considered to be selective for K_{IR}. The relative contribution of these two mechanisms may, however, differ between vascular beds and species as mice, genetically deficient Kir2.1, show a complete absence of cerebral artery dilation in response to 15 mM K^+ suggesting that only the inward rectifier mediates the K^+-mediated dilation [428]. Further, evidence has been provided for a contribution from K_{ATP} in addition to K_{IR} and the Na^+/K^+-ATPase [301]. Regardless, the basic pattern of vasodilator responsiveness to relatively low concentrations of K^+ has been confirmed in a number of in vitro and in vivo preparations, including skeletal muscle [46, 188, 255]. Activation of K_{IR} and the Na^+/K^+-ATPase results in VSM hyperpolarization, decreased intracellular Ca^{2+} and vasodilation.

Collectively, the above data provide support for K^+ as a relevant metabolic dilator. This is particularly in the case of functional hyperemia associated with muscle contraction as extracellular K^+ concentrations are known to increase to appropriate levels for vasodilation. In the case of changes in perfusion pressure (in the absence of altered metabolic activity) a role for changes in interstitial $[K^+]$ is less clear.

6.5   LACTIC ACID, CO_2 AND H^+

CO_2 and lactic acid are produced as by-products of mitochondrial respiration and anaerobic ATP production. These factors lead to an increase in H^+. During increased cellular activity, these factors increase in the interstitium and have thus been considered to be candidate factors for regulating blood flow, particularly during functional hyperemia.

Lactate has been recognized as both being released during muscle contraction and increasing blood flow for over 100 years [142]. Further, in some studies, administration of exogenous lactate can elicit a vasodilation [192]. Arguing against an important role for lactate is that (1) its levels do not correlate well with exercise intensity [256, 261]; (2) the temporal pattern of lactate accumulation does not correlate well with the temporal characteristics of the increase in blood flow [261]; and 3. lactate levels can remain increased despite reversal of the hyperemia [256, 261, 332]. Further, difficulty in understanding the exact role of lactate in local blood flow regulation lies in the fact that other metabolically relevant substances (for example, O_2 and pyruvate) alter its effectiveness as a vasodilator. Thus, lactate-induced dilation is inhibited by hypoxia [60] and increasing concentrations of pyruvate [192]. The latter observation may be relevant to the apparent dissociation between lactate levels and blood flow in the post-exercise recovery period when pyruvate levels tend to increase.

The exact mechanism by which lactate causes vasodilation remains uncertain. K^+ channel involvement, specifically BK_{Ca}, has been implicated in electrophysiological studies of smooth muscle cells isolated from coronary vessels [290]. BK_{Ca} activation may occur indirectly via a lactate-induced increase in NADH levels and production of H_2O_2. As mentioned above, hypoxia inhibits lactate-induced dilation, perhaps by decreasing superoxide and H_2O_2 production [60]. In contrast, H^+ inhibited the activity of BK_{Ca} [290] suggesting acidification did not explain the effect of lactate. Retinal arterioles dilate to lactate by a mechanism dependent on K_{ATP} channels and NOS activation [174]. In that study, uptake of lactate activated NOS with subsequent stimulation of guanylate cyclase and cGMP-dependent activation of K_{ATP} [174].

6.6 CO_2

Evidence for CO_2 acting as a local dilator has been provided by in several tissues. A difficulty in interpreting these results arises from uncertainty as to the direct effects of CO_2 as compared to indirect actions of acidification. Similarly, under in vivo conditions, changes in pCO_2 often occur in the presence of changes in other factors such as pO_2.

In the cerebral circulation, acute changes in CO_2 act as a potent dilator of pial arterioles, resulting in a 2–4% increase in blood flow for each mm Hg increase in blood pCO_2 [128]. This is attributed to the ease of movement of CO_2 across the blood brain barrier and acidification (increased $[H^+]$) of the extracellular fluid, which in turn causes dilation of the pial arterioles [410]. This mechanism is less effective during chronic hypercapnia as changes in cerebrospinal fluid bicarbonate levels buffers the effects of CO_2-induced acidification [389].

CO_2-induced increases in skeletal muscle blood flow similarly appear to be due to the associated increase in $[H^+]$. Supporting this at the level of the microcirculation, the vasodilator response to increasing pCO_2 is inhibited by concurrent superfusion of raised O_2 levels [272, 383]. In the coronary microcirculation (studied under conditions of constant myocardial O_2 consumption) an increase in pCO_2 decreases vascular resistance [160] while a decrease in pCO_2 decreases resistance [51].

Increased CO_2 has also been reported to lead to increases in other putative metabolic vasodilators including adenosine, NO and prostaglandins [50, 160, 190, 282]. Collectively, the available evidence suggests that CO_2 per se is unlikely to be the metabolic vasodilator but may exert a contributory influence through the generation of other factors.

6.7   ARACHIDONIC ACID METABOLITES

Arachidonic acid, a 20-carbon unsaturated fatty acid, can be liberated from plasma membrane phospholipids via the enzyme phospholipase A_2 or from diacylglycerol (DAG) by DAG lipase. Arachidonic acid can then be metabolized to a variety of paracrine-acting vasoactive species by cyclooxygenase, lipooxygenase or cytochrome P-450 monooxygenase. These alternate pathways give rise to prostaglandins, leukotrienes and epoxyeicosatrienoic acids (EETs). While blockade of these pathways has been associated with impaired hyperemic and autoregulatory responses, no single arachidonic acid metabolite has been conclusively shown to be a critical factor across all vascular beds. As mentioned above some metabolites of arachidonic acid may be generated under hypoxic conditions, providing a link between tissue pO_2 levels and local blood flow regulation.

The most conclusive evidence for arachidonic acid metabolites being involved in blood flow autoregulation comes from studies of Harder and colleagues in the cerebral vasculature [167]. Cerebral vascular smooth muscle cells express the cytochrome P-450 4A enzyme that catalyzes the conversion of arachidonic acid to 20-hydroxyeicosatetraenoic acid (20-HETE). This metabolite has been shown to be produced in response to increases in intraluminal pressure and is known to cause depolarization and vasoconstriction [143, 167]. Consistent with this effect, 20-HETE activates PKC and causes inhibition of BK_{Ca} [242]. These investigators have further shown that inhibition of 20-HETE formation decreases pressure-induced myogenic constriction in isolated vessel preparations and attenuates dilation of the cerebral vasculature in vivo [143, 167]. Despite this evidence, it is unclear how cerebral vessels can simultaneously utilize inhibition of BK_{Ca} to cause myogenic contraction and stimulation of BK_{Ca} to attenuate myogenic contraction [299]. The answer to this apparent discrepancy may lie in the particular cerebral vessels that were studied, compartmentalization of signaling molecules, or temporal aspects of the response.

6.8   OSMOLALITY

The osmolality of interstitial fluids increases during increased metabolic activity. Examples include increases that occur in muscle during contraction [259], in intestinal villi during digestion [32] and in glands during secretion. Consistent with a role for increases in osmolality, hyperosmotic solutions cause both vasodilation and increased blood flow in in vivo microcirculatory preparations [32, 107, 402]. Similarly, hyperosmolality causes dilation of isolated arterioles [199]. The precise role of hyperosmolality, per se, in dilation associated with increased tissue activity is difficult to assess, as a number of the accumulating osmolites may also be vasoactive. In all likelihood changes in osmolality contribute a component to the overall metabolic influence on blood flow.

The exact mechanism of how hyperosmolality affects arteriolar diameter is uncertain as both the VSM and endothelial cells may be perturbed by the changes in osmolality. Hyperosmolality has been shown to hyperpolarize endothelial cells with evidence for contributions from K_{ATP} channels and a Cl^- channel referred to as the volume regulated anion channel (VRAC). VRAC is activated by cell swelling and reduced intracellular ionic strength [184]; hence an extracellular hyperosmolality would suppress this current. The resultant endothelial cell hyperpolarization could lead to relaxation by several mechanisms including direct coupling through MEGJs; an increased driving force for the passive entry of Ca^{2+} from the extracellular space; activation of TRP channels (for example TRPV4) that are Ca^{2+} permeable. Increased $[Ca^{2+}]_i$ would lead to production of vasodilator factors (NO, prostaglandins and EDHF). Consistent with this, such paracrine factors have been reported to mediate the dilator effects of osmolites [199, 371].

Collectively, the above discussion suggests that at present a single metabolite cannot be identified as the factor underlying autoregulation of local blood flow. This does not, however, necessarily lead to the conclusion that metabolic factors are not important. In fact, it is likely that redundancy exists within the system and that metabolic control requires contributions from a number of metabolites. This, together with interactions that may occur between these factors, increases the possible complexity of metabolic mechanisms of local blood flow control. Further, current methodological limitations may impact our ability to accurately assess metabolite levels within restricted and crucial domains.

. . . .

CHAPTER 7

Endothelial-Mediated Mechanisms of Local Blood Flow Regulation

The importance of the endothelium to the regulation of arterial diameter and hence blood flow was highlighted by the discovery that this single layer of cells produces a variety of locally acting, or paracrine, factors. Initial observations related to the vascular actions of arachidonic acid metabolites [144, 285, 302] that had been described by Bergstrom and Samuelsson in seminal fluid in 1962 [25]. Among these metabolites, prostacyclin was identified as an extremely potent vasodilator and inhibitor of platelet aggregation [45]. Particularly important were the later observations of Furchgott and colleagues [140] showing that the endothelium was indeed necessary for vasodilation in response to substances such as acetylcholine and bradykinin. While the activity provided by the endothelium (denoted initially as endothelial dependent relaxation factor, EDRF) was shown to be independent of arachidonic acid metabolism, subsequent studies demonstrated that endothelial cell receptor activation led to the production of nitric oxide (NO) that diffused to the underlying smooth muscle cells to cause vasorelaxation by a mechanism involving cGMP.

More recently it has been appreciated that the endothelium also mediates vasodilation by pathway(s) involving hyperpolarization (for recent reviews, see [114, 132, 141]). Current research is aimed at determining whether this phenomenon involves production of one or more endothelial-derived hyperpolarizing factors (EDHF) or occurs by direct transfer of current between endothelial and smooth muscle cells via anatomical connections provided by MEGJs. A common factor in EDHF- and MEGJ-mediated vasodilation appears to be the involvement of Ca^{2+}-activated K^+ channels, particularly the small and intermediate conductance channels (see Table 2). Of direct relevance to local control of blood flow is the observation that hyperpolarization-mediated vasodilation appears to be more dominant in the resistance and microcirculatory vessels as compared to conduit arteries [95, 353].

Endothelial cells produce, in addition to vasodilating factors, a number of vasoconstrictor factors—some have which have been termed endothelial-derived constricting factors (EDCFs) [131, 399]. Although still to be fully characterized, EDCFs include cyclooxygenase-dependent products, reactive oxygen species and endothelin [131, 399]. Overall, an imbalance in the levels of endothelial-dependent dilator to constrictor factors has been suggested to contribute to several disorders characterized by vascular dysfunction, including states of oxidative stress, ageing, spontaneous hypertension and diabetes [399].

Direct coupling of endothelial to smooth muscle cells via MEGJs provides a pathway by which signaling molecules can potentially be shared [101, 341]. Dora and Duling suggested that direct coupling provided a mechanism by which the endothelium could modulate the extent of contraction to exogenously applied phenylephrine [100]. Thus, stimulating the α_1-adrenoceptor in smooth muscle cells of small mesenteric arteries results in an increase in intracellular Ca^{2+} in endothelial cells leading to the production of vasodilator factors that oppose the contraction. More recently, studies using mesenteric vessels from a mouse expressing an endothelial cell fluorescent Ca^{2+} sensor [382] have proposed that local signaling within the MEGJs involves the transfer of IP_3 from VSMCs to the endothelium to elicit vasodilation [246, 297]. Associated with these signaling events are discrete transient increases in Ca^{2+} that have been termed pulsars [246]. Further, these authors demonstrated the modulatory effect of the endothelium on electrically stimulated, adrenoceptor-mediated, constriction supporting its in vivo relevance [297]. Several other groups have similarly stressed the importance of restricted signaling domains, involving MEGJs, in the interactions between VSM and endothelial cells [339, 388]. Nelson's group has further proposed that endothelial cell TRPV4 channels also link to the opening of IKCa (and SKCa) through an additional local Ca^{2+} signaling mechanism that does not appear to be restricted to MEGJs and IEL holes [363]. The Ca^{2+} events in this case were termed 'sparklets' and show distinct temporal characteristics to the 'pulsars' that were described in the studies of MEGJ signaling.

As mentioned earlier, removal of the endothelium does not alter myogenic responsiveness in cannulated arterioles in the absence of luminal flow (as shown by pressure–diameter relationships and underlying levels of smooth muscle cell membrane potential), suggesting that the endothelium does not act as a direct feedback mechanism for this mode of contraction. This is possibly explained by myogenic constriction not being associated with changes in intracellular Ca^{2+} and presumably second messengers such as IP_3 that are smaller than typically seen during agonist-induced responses [434, 435].

Two specific phenomena that are characterized by involvement of the endothelium and are involved in local control of the circulation, are (1) conducted vasomotor responses and (2) flow (shear)-dependent dilation. Both of these appear to be physiologically relevant mechanisms as they can be demonstrated across species and vascular beds, and in the case of flow-dependent mechanisms, have been implicated in human vascular control [13, 69]. Further, measurements of flow-dependent (or flow-mediated) dilation have evolved to the point of being used as a prognostic indicator in the clinical study of cardiovascular pathophysiology [42, 54, 281]. These mechanisms will be discussed in the following section.

7.1   CONDUCTED VASOMOTOR RESPONSES

Conducted vasomotor responses refer to the propagation of either a vasoconstriction or a vaso-dilation from the site of stimulus (Figure 23) [13, 348]. Importantly, the vasomotor response is conducted through the cells of the artery wall and cannot be explained by simple diffusion of a chemical stimulus (for example locally applied acetylcholine) or a locally produced paracrine factor. Conduction of the underlying cellular signals is facilitated by intercellular junctions, which allow the wall to function as a syncitium. The physiological importance of this mechanism is that it al-lows for coordination of multiple elements within a vascular network (Figures 23–25). For example, if in a working muscle metabolites accumulate around the pre-capillary arterioles, dilation will be supported by the vasodilator response being conducted in a retrograde manner upstream to further lower resistance and increase blood flow. It has been argued that such coordination is required to achieve the increases in blood flow that can be observed in working skeletal muscle [348].

Conducted vasomotor responses have been experimentally demonstrated using both in vivo and in vitro preparations. The in vitro preparations have been performed both in single cannulated vessels and in branched preparations, the latter being able to show conduction beyond branch points [22]. An often used experimental approach involves the local application of a vasoactive substance (as a point source) using microiontophoresis or pressure injection through a glass micropipette. Simultaneously with application of the vasoactive agent, measurements of vessel diameter are taken at defined distances upstream of the application site (Figure 24). Diffusion of the locally applied

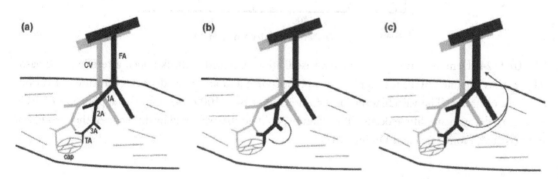

FIGURE 23: Schematic diagram showing the upstream propagation of vasodilation in skeletal muscle. Panel a depicts a feed arteriole (FA) branching into progressively small arterioles (1A, 2A and 3A) that subsequently give rise to terminal arterioles (TA) and the capillary bed. Corresponding venules are also shown. Panel b shows a vasodilator stimulus generated at the terminal arterioles which ascends to the 2A and 3A vessels to increase local flow. Panel c shows the vasodilation ascending to the feed arterioles to further amplify the increase in flow to the muscle. From Bagher and Segal [13].

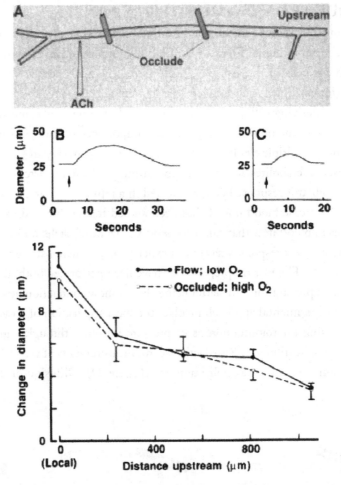

FIGURE 24: Demonstration of conducted vasodilation in hamster cheek pouch arterioles. The vaso-dilator, acetylcholine (ACh) was applied as a point source and arteriolar diameter measured at various distances upstream. The vasodilation can be observed at over 1000 μm from the application of ACh although a decay in amplitude occurs. The dilation could not be explained by diffusion of the vasodilator and persisted. From Segal and Duling [348].

agent is prevented by maintaining a continuous superfusate flow away from the site of application. Further, vasomotor responses remote to the site of stimulation have been demonstrated in response to myogenic stimuli where diffusion of an (applied) mediator would not be a consideration [325]. An alternate approach is to introduce the agonist into a side branch of an arteriole downstream of the site of observation (Figure 25).

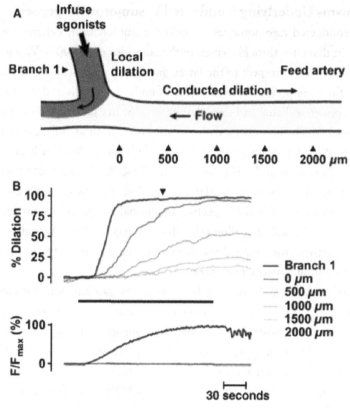

FIGURE 25: Conducted dilation in isolated and cannulated small mesenteric arteries. Conducted dilation is demonstrated following luminal perfusion of the β-adrenergic agonist, isoproterenol. (A) Isoproterenol was infused through a branch at an arterial bifurcation along with a fluorescent marker (carboxyfluorescein) to verify that the agonist did not directly enter the vessel segment used to assess conducted dilation. (B) Simultaneous traces of arterial dilation (upper panel) and relative fluorescence (F/F_{max} in Branch 1, lower panel) in response to infusion of 1 μM isoproterenol and 0.1 lM carboxyfluorescein into Branch 1. The arrowheads in A indicate the positions of diameter and fluorescence measurement in Branch 1 and upstream from the bifurcation into the feed artery (0–2000 μm). The bar indicates infusion of L-NAME (100 μM) and phenylephrine (0.5 μM) resulting in tone approximately 60% of maximum contraction. Note the delay in dilation at 0 μm compared to Branch 1. From Belezenai et al. [22].

7.1.1   Mechanisms Underlying Conducted Vasomotor Responses

A key element in conducted vasomotor responses is the gap junction, a channel with a pore of approximately 9 nm in diameter through which both molecules of <1000 MW and charge (electrotonic spread) can travel. With respect to the latter, gap junctions provide a low resistance electrical pathway. Gap junctions are present between pairs of endothelial cells and pairs of smooth muscle cells as well as between endothelial and smooth muscle cells. In each cell, 6 connexin (Cx) molecules form a hemi-channel which then docks with a hemi-channel on an adjacent cell to form a functional gap junction [396]. Although some 25 individual Cx molecules have been described, arteries and arterioles typically express and utilize only Cx37, Cx40, Cx43 and Cx45 [94]. Gap junctional function may also be regulated by post-translational regulation of the Cx molecules such as by phosphorylation [286]. From a functional perspective, functional gap junctions provide the connectivity between cells of the vascular wall that allows signals to propagate along the wall and for it to behave as a syncitial unit. Further, this connectivity allows communication between vessel branch orders possibly thereby promoting network responses.

An additional level of regulation may be provided by gap junctions forming cellular microdomains in which signaling molecules are concentrated near, or at, the pore of the junction. That this may occur has been demonstrated in studies of myoendothelial junctions where the cellular projections forming the junction appear to concentrate particular Ca^{2+}-activated K^+ channels and endoplasmic reticulum as well as exhibiting unique Ca^{2+} transients that have been termed 'pulsars' [246]. As referred to earlier, a further component of this MEGJ microdomain are the holes in the IEL through which the projections must travel to form the cellular junction.

Electrical signaling via gap junctions has been suggested to be dependent on the opening of endothelial cell plasma membrane small and intermediate conductance, Ca^{2+} activated, K^+ channels. In the case of acetylcholine, these two classes of K^+ channels are activated secondarily to a receptor-mediated increase in IP_3 production and subsequent release of Ca^{2+} from the endoplasmic reticulum. Activation of the small and intermediate conductance K^+ channels leads to hyperpolarization and spread of charge along the vessel wall. Spread of the hyperpolarization from endothelial to smooth muscle cells results in closure of VGCC and relaxation leading to dilation (see Figure 26).

In addition to conduction via electrical mechanisms, evidence has been provided for conduction of vasomotor responses by slower mechanisms such as the propagation of an intercellular Ca^{2+} wave. Such a component of conducted responses has been isolated by performing conduction studies after inhibiting small and intermediate K^+ channels [99] and using genetically bred mice with an encoded endothelial cell fluorescent Ca^{2+} sensor [382]. This raises the question of what signal is transferred from the site of stimulus to more proximal sections exhibiting vasodilation. It is unlikely that Ca^{2+} is transferred in significant amounts via gap junctions as it would be expected to quickly buffered by binding to proteins and uptake into organelles [79, 268, 388]. In contrast, IP_3 has been suggested to be transferred between cells through gap junctions [297, 388] and could, therefore,

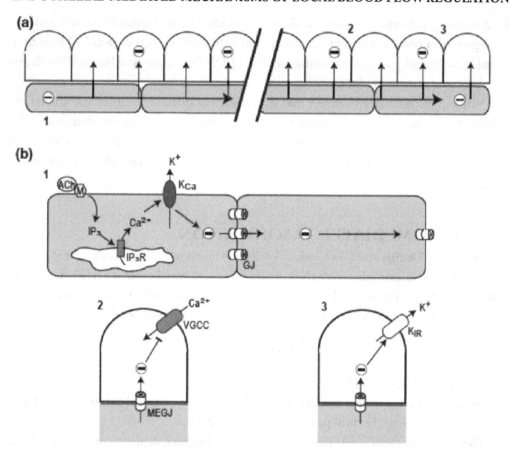

FIGURE 26: Panel a. Schematic diagram illustrating conduction of vasomotor signals along the vessel wall (electrotonic spread). ECs are shown in gray and VSMCs in white. Hyperpolarization is shown to move through gap junctions (GJs) between ECs and MEGJs between EC and VSMCs. Panel b. Signaling events hypothesized to underlie conduction. Agonist binding to EC receptors increases Ca^{2+}_i by an IP_3-dependent mechanism, which activates Ca^{2+}-sensitive K^+ channels causing hyperpolarization (1). The hyperpolarization spreads along the wall causing inhibition of VSMC VGCCs (2). Closure of VGCCs results in vasodilatation. Hyperpolarization may also be enhanced by opening of K_{iR} (3). From Bagher and Segal [13].

contribute to a regenerative response. The propagated Ca^{2+} response subsequently facilitates the release of NO and dilation.

As with many of the local regulatory mechanisms, there may, indeed, be multiple pathways for conduction. For example, de Wit [93] has suggested that while acetylcholine normally conducts its signal along the endothelium, adenosine-mediated conduction occurs between smooth muscle cells. Interestingly, genetic deletion of Cx37 and 40 resulted in conduction between smooth muscle cells suggesting a possible level of redundancy. Similarly, mediators other than NO have been implicated in conducted responses including candidate EDHFs, such as cytochrome P450-dependent arachidonic acid metabolites [186].

While it is reasonable to expect that conducted vasodilation occurs in the arterial trees of human skeletal muscle, to our knowledge, this phenomenon is yet to be directly demonstrated in the human circulation. In vivo support is, however, provided in a number of experimental studies where conduction has been interrupted chemically or via the use of mice genetically deficient for specific connexins [349, 420].

7.2   FLOW-MEDIATED VASODILATION

The flow of blood across arterial endothelial cells provides a mechanical force that stimulates the production of vasodilator substances (NO, arachidonic acid metabolites, EDHF). This mechanical force is exerted along the longitudinal axis of the blood vessel and provides a shear stress to the endothelium. Shear stress exerted on the vessel wall can be calculated as follows:

$$WSS = 4\eta Q/\pi r^3,$$

where η is the fluid viscosity, Q is flow, and r is the vessel radius.

From this equation, it is evident that shear stress is directly proportional to blood flow while being inversely proportional to the third power of the vessel radius.

Experimentally, flow- and shear-induced dilation has been demonstrated using isolated and cannulated arteriole preparations (under conditions of controlled pressure and flow) [237] and by increasing perfusate viscosity [229]. Direct in vivo evidence was provided by studies examining the effect of parallel vessel occlusion [228, 358] using microcirculatory preparations. In these latter studies, occlusion of one of a pair of daughter vessels at a branch point increases the flow (and hence shear stress) in the unoccluded branch, eliciting a vasodilator response.

The physiological relevance of flow dependent dilation is thought to be manifest as dilation of intermediate sized arterioles when smaller downstream vessels dilate in response to a stimulus (for example an increase in vasodilator metabolites). Thus, flow-dependent dilation of small arteries would 'recruit' this segment into the integrated response of the microvascular network. This is discussed further in Chapter 8, Interactions between Local Control Mechanisms.

7.2.1   Cellular Mechanisms

The mechanisms that underlie the sensing and responsiveness to flow, or specifically shear stress, are further examples of mechanotransduction within the vasculature. As described for pressure earlier, shear stress appears to exert a direct effect on the endothelial cell membrane, not involving a classical receptor-based mechanism as used by agonists. Putative endothelial cell mechanosensors/mechanosensor elements include ion channels, the membrane lipid bilayer, caveolae, glycocalyx, specific molecules such as PECAM-1, extracellular matrix proteins/integrins, inter-cellular junctions and the cytoskeleton. Additional complexity may arise from a number of these mechanosensory pathways interacting. It is likely that multiple mechanosensors may contribute to the integrated response and that this may contribute to the observed biphasic nature of shear stress responsiveness [234]. Thus, evidence has been presented for an initial phase where the onset of shear stress is detected and a secondary phase where a stable level of shear stress is achieved. Regardless of the exact mechanosensory mechanism(s), many of these pathways converge on the production of NO although other locally acting endothelium-derived paracrine factors/activities (for example prostacyclin and EDH(F)) may contribute.

While several of the candidate mechanosensory mechanisms are outlined below, the reader is referred to recent reviews for a more detailed discussion [8, 18].

7.2.2   Ion Channels

A number of endothelial cell ion channels have been demonstrated to exhibit mechanosensitivity. In particular, several classes of K^+ channels have been implicated in the shear stress response. For example, Olesen et al. identified a shear-activated K^+ current that led to membrane hyperpolarization that reversed on removal of the stimulus [306]. Evidence was later provided for specific involvement of Kir2.1 [187].

A possible functional consequence of endothelial cell membrane hyperpolarization is entry of Ca^{2+} via non-selective cation channels. The resulting increase in Ca^{2+} would favor activation of eNOS and production of NO. At present, little of this evidence comes from in situ or freshly isolated microvascular endothelial cells; further Ca^{2+}-dependent mechanisms of NO production do not appear sufficient to, alone, explain shear-dependent dilation [18].

More recently, interest has focused on the role of several TRP channel proteins in shear-dependent responses; including TRP V4, C1, 4 and 6 and polycystins [171, 224, 260, 429]. A role for the Ca^{2+}-permeable non-selective cation channel, TRPV4, in flow-dependent dilation has been demonstrated both pharmacologically [224] and using genetically deficient mice [171, 280]. Currently, it is believed that TRPV4 is not, itself, mechanosensitive but requires stimulation by epoxy-eicosatrienoic acid metabolites of arachidonic acid [429].

If there is a primary role for ion channels in shear-dependent responses, an obvious question relates to how ion channels are activated by the mechanical stimulus. Similar arguments have been made regarding the role of ion channels in myogenic signaling (see Figure 22) including a direct change in ion channel protein conformation, changes in channel protein conformation secondary to effects on the lipid bilayer and via tethering mechanisms to either the ECM or cytoskeletal proteins.

7.2.3   Glycocalyx

The luminal surface of endothelial cells is coated with a filamentous glycoprotein layer projecting from the cell membrane into the vessel lumen. Enzymatic (hyaluronidase, heparinase, neuraminidase) disruption of the glycocalyx results in marked attenuation of the shear stress-mediated production of NO [283, 313, 317]. While the mechanism of how shear stress initiates glycocalyx-mediated mechanical signaling is uncertain, a conformational change in this layer may transmit signals intracellularly to the actin cytoskeleton [80, 411]. Alternately, the glycocalyx has been suggested to affect mechanosensitivity through the production of reactive oxygen species [108]. Interestingly, H_2O_2 peroxide has been proposed to mediate flow-dependent dilation in human coronary arteries via activation of BK_{Ca} channels [252].

7.2.4   Caveolae

As mentioned earlier, caveolae are cholesterol-rich microdomains which appear as flask-shaped invaginations (50–100 nm across) in the surface of plasma membranes. Caveolae and their associated caveolin proteins localize a number of ion channels, receptors, kinases and enzymes in signaling complexes. In particular, endothelial nitric oxide synthase (eNOS) is bound to caveolin from where it can be released by increases in Ca^{2+}_i and activated by phosphorylation. Supporting a role for caveolae/caveolin 1 (Cav-1) in endothelial cell mechanotransduction, Cav-1-/- knockout mice show impaired flow dependent dilation and flow-induced remodeling [427]. Further, caveolae have been implicated in flow-induced activation of signaling pathways including tyrosine phosphorylation events and the activation of eNOS [326, 327].

In addition to caveolae, membranes contain other lipid-rich microdomains (for example, lipid rafts). These domains may also be impacted by shear forces [380].

7.2.5   G-Protein Activation

Shear stress-mediated activation of trimeric G-proteins has been shown in model membrane systems (lipid bilayer) [157] and in several cell types [158]. Evidence exists for shear exerting direct activating effects on the G-protein complex [157] or by changing the conformation of a G-protein-coupled receptor (B2 bradykinin receptor [55]), which subsequently causes activation of the G-protein. Importantly, in both of these putative mechanisms, G-protein activation occurs independently of the binding of a classical agonist. Downstream of G-protein activation is the activation of phospholipases and subsequent generation of second messengers such as DAG, PKC, IP_3 and Ca^{2+}. Conceivably, these factors would link to events including modulation of ion channel activity and Ca^{2+}-dependent enzymes.

7.2.6   Platelet-Endothelial Cell Adhesion Molecule 1 (PECAM-1)

PECAM is a specific cellular adhesion molecule that has been implicated in the detection of shear stress by endothelial cells [307]. PECAM localizes itself to cell–cell junctions and associates with a PECAM molecule from the adjoining cell. Interestingly, PECAM forms part of a possible mechanosensory complex also being associated with integrins, VE-cadherin and VEGF receptor protein all of which have been implicated in endothelial cell detection of shear stress. Consistent with a critical role for this adhesion molecule, studies in PECAM knockout mice have shown impaired flow-dependent dilation in arterioles from skeletal muscle and coronary arteries [14, 253].

7.2.7   Integrins

Integrins are an attractive mechanosensor as they can be mechanically activated and link through focal adhesions to the cytoskeleton and a number of signaling pathways (for example, involving tyrosine phosphorylation such as in the activation of focal adhesion, cSRC and MAP kinases). Activation of integrins by shear may occur at the luminal surface or at the abluminal surface as a result of coupling through the cytoskeleton. Supporting a role for integrins, synthetic RGD containing peptides (tri-peptide sequence found in ECM proteins) inhibits flow-dependent dilation in isolated porcine coronary arterioles [291].

Thus, multiple pathways appear to be involved in the detection of, and responses to, changes in shear stress (Figure 27). This complexity appears similar to how vascular cells respond to other mechanical stimuli including changes in intraluminal pressure and membrane tension. Also similar is that alterations in shear stress, in addition to stimulating the acute production of vasodilator factors to affect changes in vascular tone, are linked in the longer term to gene expression (for example, modulating eNOS at the transcriptional level). In this way, local blood flow regulation can be further coupled to vascular remodeling events.

7.3   CLINICAL MEASUREMENTS OF FLOW DEPENDENT DILATION

Although not performed as a true measure of microvascular function, brachial artery flow-mediated dilation (FMD) is being studied as an index of endothelial dysfunction [54] and as a prognostic indicator of future vascular events [42, 281]. In this procedure the diameter of the brachial artery

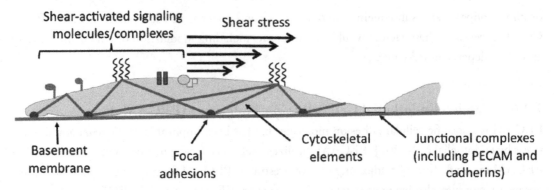

FIGURE 27: Schematic diagram illustrating hypothetical mechanisms by which blood flow may affect mechanotransduction in endothelial cells. As described in the text, changes in shear stress may be detected by cell surface-associated signaling molecules/complexes (ion channels, G-proteins, mechanosensitive enzymes) and/or junctional complexes (cell-cell and cell-matrix).

is measured using ultrasound before and after the release of an occlusion created by inflating a sphygmomanometer cuff. Releasing the cuff stimulates a reactive hyperemia with the increase in blood flow increasing shear stress on the brachial artery endothelial cells leading to local vasodilation. The relaxation is believed to be largely a result of NO release although this may depend on the exact methodological approach used (particularly considering the site of the occlusion cuff [153]) and contributions may also be provided by other endothelium-derived prostaglandins and/or EDHFs [372].

. . . .

CHAPTER 8

Interactions Between Local Control Mechanisms

In the preceding chapters, the various local blood flow control mechanisms have been discussed in isolation. In this chapter, we take a more integrated approach considering two well-studied situations in which all of these mechanisms likely participate—either simultaneously or during phases of the response. Further, a simplified model of series coupling of these mechanisms is considered [238]. In this model, segments of the circulation are considered as being largely specialized to respond to metabolic, myogenic or flow-dependent stimuli.

8.1   REACTIVE HYPEREMIA

Following the release of a transient occlusion of an artery, a period of increased blood flow (relative to that prior to the occlusion) can be measured (Figures 28–30). This increase in flow, termed *reactive hyperemia,* has been viewed as being necessary to repay the metabolic debt accrued during the transient ischemic period and/or to provide rapid removal of accumulated metabolites. As the magnitude of the hyperemic response has been shown to be directly related to the duration of occlusion (Figures 28–30) [208], it has been assumed that accumulation of tissue-derived metabolic factors underlies the dilation of arterioles occurring after release of the occlusions. A number of metabolic factors, described in the previous chapter, have been suggested to be involved in reactive hyperemia including adenosine, K^+, prostaglandins and nitric oxide. In addition to an accumulation of metabolites, the decrease in arterial pO_2 has been implicated in the subsequent hyperemic response, although several studies (both at the whole organ and microvascular level) have shown that reactive hyperemia can be observed when an occlusion-release protocol is performed under conditions in which pO_2 levels are maintained [230, 254]. In contrast to this, Tuma et al. [390] did observe a decrease in the extent and duration of hyperemia at elevated pO_2. While this may reflect some species or vascular bed differences, it is evident that at least a degree of reactive hyperemia can still be demonstrated when pO_2 is maintained thus implicating the involvement of factors other than pO_2.

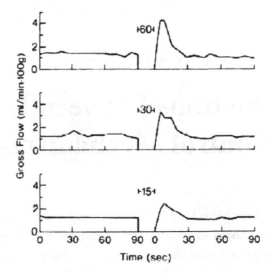

FIGURE 28: Reactive hyperemia responses to increasing duration of occlusion (15–60 seconds). Note the increasing amplitude of the post-release hyperemic response. Experiments were conducted in dennervated cat sartorius muscle. From reference [208].

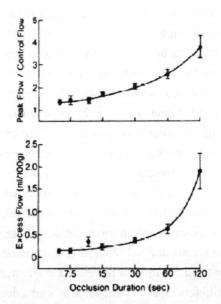

FIGURE 29: Quantification of the extent of reactive hyperemia with increasing occlusion duration. As in Figure 28 experiments were conducted in cat sartorius muscle. From reference [208].

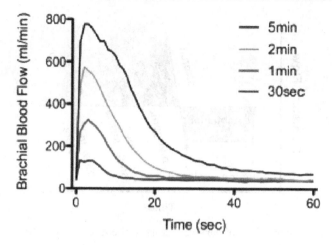

FIGURE 30: Duration and magnitude of reactive hyperemia is related to the length of occlusion. The figure shows the increase in brachial artery blood flow (human subjects) following the release of an occlusion (30 seconds–5 min duration) applied by inflation of a sphygmomanometer cuff. Blood flow was measured by Doppler ultrasound. From Clifford [69].

Reactive hyperemia can be demonstrated at the organ level by occluding inflow to the organ. Similarly, it can be shown in microcirculatory preparations where a blunt occluder pipette is typically pressed against a single arteriole to stop blood flow. Reactive hyperemia of arterioles can also be observed after release of a venous occlusion; however, in this situation, pressure is increased on the arteriolar side prior to restoration of blood flow. Importantly, reactive hyperemia can be demonstrated in human subjects as well as in experimental animal preparations. A common approach used in the study of human subjects involves inflation of a sphygmomanometer cuff to occlude blood flow. Resultant changes in flow can be measured by a variety of techniques including Doppler ultrasound, laser Doppler flow probes or capillaroscopy.

A number of observations suggest that the hyperemic response cannot be simply explained by a single metabolite. For example, the response varies in magnitude between tissues, very brief occlusions (assumed to be too short for significant accumulation of metabolites) have been shown to elicit dilation on release and can be reproduced (to a certain extent) in vitro in the absence of parenchymal tissues (presumed source of metabolites) [227]. This is perhaps not surprising when one considers that in addition to the possible accumulation of metabolites, pressure and shear stress change markedly from the occluded to the re-perfused state. Changes in pressure, in particular, may cause both passive changes in diameter and active myogenic responses. An increase in flow following release of the occlusion would be expected to lead to a shear stress-induced release of the vasodilator, NO.

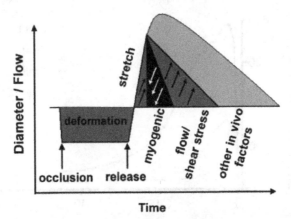

FIGURE 31: Proposed contributions of various factors to the reactive hyperemia following release of a transient vascular occlusion. The concepts were, in part, developed from in vitro studies of isolated cannulated arterioles in the presence and absence of flow. These authors propose that reactive hyperemia results from a combination of mechanical (cell deformation, pressure and shear-induced) and metabolic factors (the latter denoted as 'other in vivo factors in this figure.' From Koller and Bagi [227].

To isolate the contributions of these components (particularly those resulting from mechanical factors), Koller and Bagi [227] studied isolated and pressurized rat gracilis arterioles in the absence and presence of flow. Changes in pressure were applied to mimic that observed during in vivo studies of vascular occlusion. This in vitro study indeed provides data to support the overall hyperemic response being, in part, a function of mechanical changes that occur during the occlusion and on restoration of blood flow—specifically changes in pressure and shear stress (Figure 31). Using the NO synthase inhibitor, L-NAME, and disruption of the endothelial cell layer an important role was established for NO and that this contribution increased with the duration of occlusion. Further, this component was increased in preparations with intraluminal flow implicating a role for shear stress in the endothelial-dependent component. Although in this preparation, NO appears to explain the endothelial cell-dependent component, additional mediators have implicated in other preparations [226].

8.2 FUNCTIONAL HYPEREMIA
The term functional hyperemia refers to the increase in blood flow that occurs to meet the increased metabolic demands during tissue activity. This may be in exercising skeletal muscle (see Exercise

Hyperemia below); the secretory glands and smooth muscle of the gastrointestinal tract during digestion; or in cerebral tissue during thought and computational processing.

8.3 EXERCISE HYPEREMIA

A common example of functional hyperemia is the marked increase in blood flow that occurs in response to physical exercise. The increase in blood flow that occurs in skeletal muscle during exercise represents one of the largest demands on the circulatory system. As a fraction of cardiac output active skeletal muscle can increase its blood flow by a factor of 20 fold. This phenomenon of activity-induced hyperemia in muscle has been appreciated since the 1870s. Despite uncertainty in the exact cellular mechanisms mediating this hyperemia the increase in blood flow is largely met by reduction in vascular resistance with MAP remaining relatively constant, despite marked variations in exercise intensity [244].

During exercise, the steady-state increase in blood flow occurs linearly with an increased metabolic/O_2 demand of the working muscle, and thus represents a local control phenomenon [244]. However, when temporal aspects of the blood flow response are considered, it is apparent that additional factors contribute. For example, an initial increase in flow (within one second of the onset of contraction) can be demonstrated prior to an actual increase in metabolism and at a time when an early decrease in a-v O_2 difference is evident. While the exact mechanisms contributing to these events are uncertain, it has been suggested that mechanical factors (for example compression) and neural pathways may be involved [70, 244]. The contribution of neural mechanisms has been debated, however, as the exercise-induced increase in flow can be demonstrated in the presence of ganglionic blockade.

A considerable body of evidence suggests that the steady-state exercise-induced hyperemia is mediated through the production of vasoactive metabolites. Tissue activity increases the accumulation of such metabolites that decrease the basal level of tone and increase vascular conductance. As described in earlier sections, the basal tone in skeletal muscle arterioles is largely myogenic in nature. The exact metabolites leading to hyperemia are uncertain, in part, because it is likely that multiple substances are involved. The apparent redundancy may indicate that the system is designed to increase blood flow despite variation in exercise type and intensity.

An important consideration in understanding exercise hyperemia is how the vasculature recruits both small and larger arterioles to allow flow to increase the very high levels observed under maximal activity. The accumulation of metabolites would likely only directly affect vessels close to the capillary network requiring mechanisms to extend the vasodilation to the larger upstream arterioles and feed vessels. Further, for metabolic dilation to be an effective local control mechanism, it would be expected to be active on single muscle fiber level and summate with increasing intensity of activity. Putative mechanisms contributing to the dilation of the proximal vessels include flow or shear stress-mediated vasodilation and propagated/conducted responses. As mentioned in earlier chapters, the former has been suggested to involve shear-induced production of paracrine factors and the latter the spread of hyperpolarizing signals via gap junctions [13].

Importantly, blood flow changes during exercise are not limited to those occurring in skeletal muscle. Coronary blood flow increases as the work load on the heart increases while skin blood flow increases in response to thermoregulatory needs. Renal and splanchnic blood flow tend to decrease as a result of neurally (sympathetic)-mediated mechanisms which in part allow redistribution of cardiac output to the working muscles.

8.4   SERIES-COUPLED MECHANISMS OF LOCAL BLOOD FLOW CONTROL

In the earlier chapter on myogenic signaling reference was made to studies showing that the intensity of myogenic contraction, or degree of myogenic responsiveness, increased as arteriolar diameter decreased. An exception to this was that immediately pre-capillary arterioles showed a relative decrease in myogenic responsiveness compared to the proximal vessels. It has, therefore, been argued that the pre-capillary arterioles may be more responsive to metabolic stimuli than to intraluminal pressure. It has been suggested that a consequence of this is there is a 'zone' or segment of the microvascular bed that can broadly be considered as the myogenic element [238]. Important questions are, therefore, do other segments of the microvasculature exhibit a propensity for particular regulatory mechanisms and, if so, how do these segments interact to give an integrated vasomotor response?

Before considering the sensitivity of particular vascular segments to specific local control mechanisms, it should be mentioned that these zones of specialization are relative (as vessels will likely exhibit multiple behaviors) and not necessarily anatomically fixed. In regard to the latter, for example, neurally mediated vasoconstriction can induce (or increase) myogenic reactivity in larger vessels while pathophysiological states, such as hypertension, can alter myogenic gain [124, 251].

In regard to sensitivity to candidate vasodilator metabolites evidence supports an increasing sensitivity as vessel diameter decreases. This has been shown by examining responsiveness to topically applied vasodilators in vivo microcirculatory preparations. Interpretation of these data can be difficult, however, as the approach may also cause changes in flow (and hence shear stress) and pressure. Kuo et al. [239] addressed this difficulty by studying the responsiveness of four branch orders (diameters approximately 40–180 μm) of coronary arterioles to exogenous adenosine under isolated and isobaric conditions. Across this range of diameters, there was an approximate two log order difference in sensitivity to adenosine with the smallest vessels showing the greatest sensitivity.

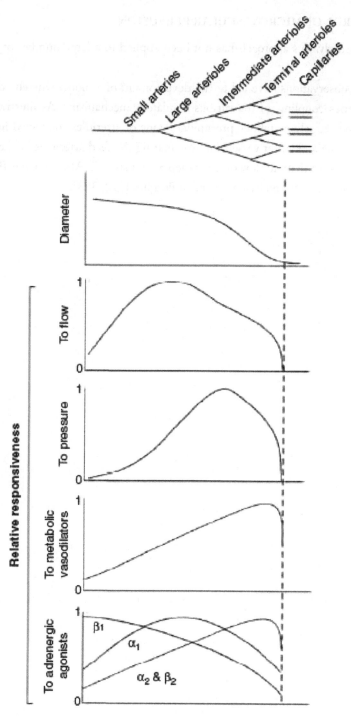

FIGURE 32: Series-coupled model for the control of microvascular blood flow. The figure illustrates the relative responsiveness of various segments of the microvascular network to metabolic, pressure, shear stress and adrenergic stimuli. From reference [88].

Somewhat surprisingly, this approach has not been applied to a large number of candidate metabolic factors.

The above observations have led to the development of a model whereby the microvasculature exhibits a 'series-coupling' of the various regulatory mechanisms. As illustrated in Figure 32 this model depicts a largely metabolic precapillary zone of arterioles connected in series to an upstream myogenic zone and larger vessels (approximating the feed arterioles in skeletal muscle) exhibiting the highest sensitivity to flow or shear-dependent stimuli. Also shown is the distribution of adrenergic receptors as described by Faber and colleagues [272, 383].

. . . .

CHAPTER 9

Interactions with Extrinsic Control Mechanisms

An important consideration is how local mechanisms of blood flow control interact with extrinsic mechanisms. As mentioned earlier, arteriolar endothelial and VSM cells possess receptors for many circulating hormones and neurally released factors. Further, arterioles and resistance vessels exist within a vascular network where the more remote direct activity of extrinsic mechanisms can alter pressure and flow that may lead to the secondary alterations in myogenic, metabolic or shear-dependent mechanisms. An example of the latter was demonstrated in vivo in the rat, where infusion of either angiotensin II or phenylephrine caused an increased in vascular resistance that was due to both the direct effect of the agonist and the increase in blood pressure [278]. The pressure-dependent, or autoregulatory component was discerned by examining the response to the agonist when arterial pressure was free to change and when the hindquarters were held at constant pressure by graded occlusion of the sacral aorta. Overall, it was concluded that the interaction between the neurohumoral and autoregulatory responses led to an 'amplified rise in vascular resistance' [278].

An additional illustration of the interaction between local and extrinsic regulatory mechanisms is the roles of the sympathetic nervous system and metabolic vasodilation during exercise. It has been argued that some degree of maintained sympathetic control is required to prevent possible decreases in MAP, secondary to an inability to further increase cardiac output, during intense exercise involving large muscle mass. Factors contributing to this integrated control include a greater adrenergic receptor (particularly α_1-adrenoceptors) density in the larger feed arterioles relative to pre-capillary vessels and a relative increase in sympatholysis as arteriolar diameter decreases [48, 272, 383].

As an example, adrenergic receptor stimulation enhances myogenic vasoconstriction in some vascular beds through downstream interaction of signaling mechanisms [251, 275]. Conversely, the prevailing level of myogenic tone often influences the effectiveness of these other regulatory mechanisms. As outlined above, this latter point has been clearly illustrated in in vivo studies of the hindquarter and splanchnic circulations of rats. In these studies, when either of the vasoconstrictors angiotensin II or phenylephrine were infused to induce systemic hypertension, the local increases in hindquarter or splanchnic vascular resistance could be markedly attenuated by protecting the vascular beds from the pressure increase [278]. Thus, despite the presence of the circulating agonist, in the absence of an additional pressure stimulus, the vasoconstriction was significantly reduced. It was concluded that pressure-dependent autoregulatory events are activated during agonist-induced contractile activation such that the myogenic mechanism responds to the changes in vascular pressure that result from the direct receptor-mediated response to the agonist. Thus, the net effects on local vascular resistance were a composite effect of both neurohumoral and myogenic components.

A consequence of these interactions is that it is important to consider the preparations used in conducting studies into microvascular function and the underlying cellular mechanisms. Thus, an isolated vessel lacks the neurohumoral and paracrine factor milieu, which (while not necessarily causing overt changes in vessel diameter under resting or basal conditions) might modulate the sensitivity to other mechanisms, for example, those involved in mechanotransduction. Similarly, an isolated cell lacks the normal mechanical forces and environmental contacts that are present in the complex 3D vascular structure. On the other hand, in vivo preparations are complex and studies of pharmacological responses are difficult to interpret when a given agent may be acting directly on the vessel under study, via remotely initiated hemodynamic actions or on parenchymal tissues.

. . . .

CHAPTER 10

Vascular Heterogeneity

Considerable variability exists with respect to the autoregulatory capacity of the various tissues (see Chapter 5). The highest level of intrinsic autoregulatory gain is observed in the cerebral and coronary vasculatures while the lowest level is observed in the skin. The highest level of autoregulatory gain is therefore seen in those tissues least dependent on the sympathetic nervous system for generation of tone while the lowest levels of autoregulatory gain are observed in the skin, which is highly dependent on sympathetic innervation. Skeletal muscle and renal vasculatures show intermediate autoregulatory gain as well as being substantially influenced by the sympathetic nervous system (Table 6; Figure 33).

Although a topic of its own, it is important to consider that tissue, network and cellular heterogeneity likely impacts mechanotransduction processes in small arteries. Presumably, this reflects both differences in the local mechanical environment and tissue function. Similarly, responsiveness to chemical stimuli is affected by local levels of the vasoactive mediator and receptor properties, including receptor density and affinity.

Within a given vascular bed or tissue, there is considerable variability in the degree of myogenic reactivity, or gain, exhibited by differing branch orders of arterioles/small arteries (see Chapter 5: Myogenic Vasoconstriction and Dilation). This has been demonstrated using both in vivo and in vitro preparations. In general, myogenic reactivity tends to increase as diameter decreases, although in some tissues, the smallest pre-capillary arterioles may be more adapted to respond to metabolic stimuli as opposed to changes in intraluminal pressure.

Recent studies utilizing confocal and multiphoton microscopy have highlighted structural differences in the arteriolar wall between vascular beds. In particular, marked differences exist in the extracellular matrix components both at the level of the internal elastic lamina (IEL) and the adventitial surface. For example, the size and architecture of IEL holes (presumed sites of myoendothelial coupling) differ between vessels. Further, while cremaster muscle arterioles and small mesenteric arteries exhibit an adventitial layer of elastin fibers, these are lacking in small cerebral arteries [71]. How these differences in structure impact myogenic signaling is currently uncertain. However, given the role of the ECM in determining the mechanical characteristics of the vessel wall and the possible role for the ECM as an element of a putative mechanosensory mechanism, it is conceivable

TABLE 6: Inverse relationship between the ability of various tissues to exhibit autoregulatory and sympathetic control.

TISSUE	AUTOREGULATORY GAIN	SYMPATHETIC INNERVATION	COMMENTS
Brain	++++	+	Allows organs vital for survival to respond to their metabolic requirements while when appropriate blood flow to other organs can be markedly diminished.
Heart	++++	+	As above
Kidney	+++	++	
Splanchnic	+++	+++	Prolonged sympathetic stimulation of arterioles leads to autoregulatory escape
Skeletal muscle	+++	++	
Glandular			
Skin	+	++++	Adapted to meet the thermo-regulatory role of the cutane-ous circulation.
Pulmonary circulation	−		

that a differing structural environment influences the ability of smooth muscle cells to both sense and responds to mechanical stimuli.

At the cellular level, heterogeneity is seen in the expression and function of important signaling molecules including ion channels. For example, evidence exists for differences in the molecular composition of BK_{Ca} such that it is more sensitive to a given change in intracellular Ca^{2+} in cerebral

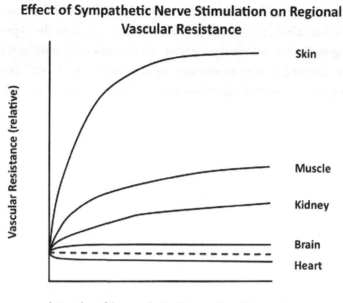

FIGURE 33: Schematic representation of the response of various vascular beds to sympathetic nerve stimulation. Relative changes in vascular resistance are drawn for increasing intensity of sympathetic nerve stimulation. Note that tissues showing the highest degrees of autoregulation show the lowest responses to sympathetic stimulation.

artery SMCs compared to those of arterioles from skeletal muscle [426]. This may account for differences in the characteristics of the pressure–diameter relationships between these two vascular beds [232]. Similarly, and as mentioned earlier, the expression of VGCC has been reported to differ between afferent and efferent arterioles of the glomerulus [258]. Heterogeneity may also exist between vascular SMCs, per se. Consistent with this, Chilton et al reported marked differences in cellular capacitance indicating that SMCs from afferent arterioles were considerably smaller than those of cerebral and cremaster muscle vascular SMCs (contrast references [64] and [426]). This group also reported heterogeneity in regard to contractile protein isoform expression such that the renal afferent arteriole predominantly expresses the MHC-B isoform, as opposed to the efferent arteriole expressing only the slower-cycling MHC-A isoform [354]. Such differences may well impact myogenic properties, for example, the kinetics of contraction, through differences in Ca^{2+} handling and contractile protein interactions.

Regional differences in cellular phenotype are not limited to vascular smooth muscle but are also seen in endothelial cells [62, 315, 356]. Such differences include the expression of differing patterns of signaling molecules, adhesion molecules and pro-atherosclerotic factors. This may relate to inherent cellular differences or responsiveness to regional environmental factors (for example, variation in levels of shear stress, turbulent flow and intraluminal pressure).

·   ·   ·   ·

CHAPTER 11

Alterations in Microvascular Blood Flow Control in Pathophysiology

Given that the microvasculature is responsible for the delivery of nutrients and removal of many metabolic products from all tissues, it is not surprising that microvascular dysfunction is a common feature of many disease states. While it is not possible to comprehensively review all such states, the following sections provide a brief insight into common disorders associated with microvascular dysfunction.

A number of techniques are now available for studying the microcirculation in human subjects and therefore in relevant pathophysiological situations. These include nailfold capillaroscopy [122, 386, 387]; retinal vascular photography (including with fluorescent tracers) [127, 216, 320, 385]; laser Doppler flowmetry [368, 369]; flow-mediated dilation [54]; and ultrasound [78, 156] Further, reactivity studies are being performed on small arteries isolated from punch biopsies (typically subcutaneous vessels) and at surgery (for example, intestinal mesentery and atrial appendage) [56, 342]. Vessels under these conditions have also been used for cell isolation for the study of smooth muscle cell ion channel function.

Disturbances in local blood flow control occur in many pathophysiological situations, including those where the primary disorder impacts the cardiovascular system and others where vascular complications are a secondary event. A common example of the latter occurs in diabetes mellitus and related metabolic disorders, where clinically relevant microvascular dysfunction is observed in the kidneys, retina and nerves [43, 44]. A clear distinction of cause and effect is not, however, always apparent as microvascular dysfunction can contribute to the development or exacerbation of a particular disorder. For example, microvascular dysfunction in skeletal muscle may in fact contribute to insulin resistance and glucose uptake thus contributing to metabolic dysfunction and the progression towards overt type II diabetes. Similarly, microvascular dysfunction can both contribute to and result from hypertension. Thus, in hypertension functional and structural rarefaction, as well as increased vascular reactivity, can contribute to the increased vascular resistance while hypertension can induce vascular remodeling at the resistance vessel level.

An important question in understanding the etiology of microvascular dysfunction in disease states, and ultimately directing therapeutic approaches, relates to whether control mechanisms are specifically impaired or overridden by environmental factors. Further, altered microvascular function may result from compensatory remodeling events as for example occurs in hypertension (increased wall: lumen ratio) or chronic ischemia (angiogenesis, collateral formation). Pathophysiological changes in microvascular reactivity may result from a variety of mechanisms including changes in receptors, ion channels, Ca^{2+} handling or at the contractile protein level.

Altered arteriolar myogenic reactivity has been reported in a number of vascular disorders including subarachnoid hemorrhage, cardiomyopathy, hypertension, obesity and diabetes [180, 249]. Thus, following subarachnoid hemorrhage, small arteries develop increased myogenic tone, which has been suggested to be due to a number of factors from closure of K_{Ca} channels in response to increased 20-HETE production [331], decreased Ca^{2+} spark activity [198, 412], an increase in Ca^{2+}-sensitivity as a result of increased Rho-kinase activity [355] and the opening of R type Ca^{2+}-channels [198]. While these states largely represent enhanced myogenic reactivity, situations of decreased myogenic constriction may contribute to pathophysiology. For example, an increase in BK_{Ca} channel expression, and an enhanced coupling of Ca^{2+} sparks to BK_{Ca}, has been implicated in decreased vascular resistance and hypotension that is evident in septic shock [432].

Endothelial dysfunction has been implicated as an early event in many vascular disorders. Of relevance to the current discussion is whether endothelial dysfunction in the microvasculature leads to specific impairment of local blood flow regulatory mechanisms. From the earlier chapters this would likely present as alterations in metabolic regulation, flow-dependent dilation and reactive hyperemia (that is, where mechanical or paracrine stimuli demonstrate dependence on the endothelium). Indeed, clinical evidence exists for reactive hyperemia being impaired in diabetic subjects and in subjects at risk for the development of diabetes [200]. Similarly, endothelial dysfunction has been reported to precede the development of hypertension in predisposed human subjects [9].

Microvascular dysfunction can also result from environmental factors. In this regard, cigarette smoking is associated with acute impairment of endothelial-dependent dilation and capillary recruitment (measured in skin) [193] possibly through the production of reactive oxygen species and oxidative stress [214]. Similarly, particulate matter, as possibly provided by environmental pollutants, while taken in through the respiratory system, has been shown to induce systemic microvascular dysfunction [305].

Further examples include dysfunction associated with disorders that involve angiogenesis. Microvascular abnormalities are seen in tumor vasculatures, hypoxia in newborns and proliferative retinopathy in diabetes. In particular, these situations are associated with abnormal growth patterns and dysfunction particularly in regard to vessel permeability properties [49, 202].

An important question is: are there situations in which specific pharmacological modulation of myogenic tone and perhaps other local regulatory mechanisms would confer an advantage over an intervention aimed at general smooth muscle cell activation/inactivation? An ability to set a new level of vascular resistance through a specific modulation of myogenic mechanisms might enable normal neurohumoral mechanisms to interact with a pharmacologically manipulated level of arteriolar tone. Such interactions are essential to the preservation of in vivo vasomotor responses, suggesting that it might be possible to specifically target, and increase, myogenic tone. This would be advantageous in situations in which the cardiovascular system is depressed (for example shock and low flow states). It is also possible that the ability to decrease myogenic tone in certain hyperdynamic states would be beneficial. To make this concept practical, increased knowledge of the cellular mechanisms that underlie local regulation of the microcirculation is required. Ultimately, this strategy may identify specific targets (including, mechanosensors, ion channels, signal transduction pathways) for pharmacological intervention.

*   *   *   *

CHAPTER 12

Long-Term Regulation of Blood Flow

The microcirculation is not a static structure and can remodel over a longer term to meet changing demands for perfusion. Remodeling occurs at the cellular level where VSM and endothelial cells adapt by altering the expression of molecules involved in the regulation of vessel diameter (for example, ion channels, Ca^{2+} signaling molecules, integrin-mediated attachment [28]). At the other, extreme hypertrophic and hyperplastic responses occur in the arteriolar wall as well as situations where there is an alteration in vessel number (vasculogenesis or rarefaction [57, 324]). A chronic alteration in perfusion demand can occur under physiological (for example exercise and exposure to high altitude) and pathophysiological (for example hypertension, chronic ischemia and diabetes) conditions. Such adaptations occur over a variety of time frames ranging from hours through a period of many days to years.

A highly studied situation of vascular remodeling is that occurring in hypertension. A characteristic of human hypertension and many experimental animal models is inward eutrophic remodeling of resistance arteries [3, 292]. That is, resistance vessels remodel to have a smaller diameter in the absence of a change in wall area. This was originally proposed by Folkow (reviewed in reference [135]) to occur in human essential hypertension and later directly by Aalkjaer et al. in small vessels from gluteal muscle biopsies [3]. Rizzoni later made the important observation that remodeling of small arteries in hypertension was predictive of the later development of abnormalities seen in established hypertension (including left ventricular hypertrophy and the occurrence of strokes and myocardial infarction) [328, 329]. The cellular and molecular events underlying such remodeling are subjects of extensive current research, with the hope of identifying pharmacological approaches for intervening in this remodeling event. Debate continues, however, as to the exact role of the small artery remodeling in hypertension—is it a cause of hypertension or, in part, an adaptive response?

Remodeling of the vasculature occurs during exercise training to meet the requirements of increased muscle mass and enhanced performance. In addition, it has been argued that the vascular adaptation to exercise is directly anti-atherogenic, providing a mechanism for the observed beneficial effects of exercise on cardiovascular outcomes. The increase in blood flow in active muscle (both cardiac and skeletal) is believed to cause alterations in gene expression through shear stress-dependent mechanisms. The effect of shear stress can be demonstrated experimentally, for example, by surgical insertion of arteriovenous fistula. Further, the effects of shear stress can be demonstrated in tissue culture models. Among the changes stimulated by shear stress are an increase in NO synthase expression and a decrease in endothelin expression. This would be expected to shift the balance from vasoconstriction to vasodilation. Further, chronic increases in shear stress lead to an outward remodeling of arterial vessels. The benefits of exercise, however, extend to vessels not directly subjected to marked increases in shear stress suggesting that other effects occur at the systemic level.

Importantly, the shear stress-dependent modulation of vascular structure can be demonstrated to function in both directions; that is, adaptations to both increases and decreases of shear stress. Thus, Langille and O'Donnell demonstrated that a 70% reduction in blood flow in the carotid artery of rabbits caused an approximate 20% decrease in vessel diameter [243]. This response was further shown to be dependent on the endothelium. In addition to overt changes in vessel dimensions vascular adaptation occurs at more subtle levels. Martinez-Lemus et al. suggest that there is a continuum of responsiveness from acute vasomotor responses to structural adaptation [267]. Thus, within four hours of inducing contraction of arterioles with noradrenaline, smooth muscle cells were shown to re-position themselves within the vessel wall [266]. This allowed the cells to re-lengthen but maintain a contracted diameter by sliding over one and other. Presumably, this cellular remodeling involves disruption of existing cell–ECM connections and likely alterations in the cytoskeleton [265]. Similar observations have been reported by Bakker and colleagues, further implicating the ECM-modifying enzyme transglutaminase [15, 16].

The above are only a few examples of adaptation/remodeling of the vasculature as a means for meeting altered requirements for blood flow. Other situations where marked changes in vascularity (with alterations at the level of the resistance vessels) occur are during pregnancy [263, 311], tumorigenesis and tissue growth and development.

Collectively, these longer-term mechanisms provide structural adaptation and contribute to local blood flow control by alterations in vessel diameter and number and in some cases through changes in the character of the vessel wall. Some of these processes represent physiological responses while in some cases they contribute to the pathophysiology of a disease.

. . . .

Summary

Local mechanisms of blood flow control allow tissues to regulate their own hemodynamic conditions so as to meet metabolic requirements while also providing a level of protection against inappropriate hypo- or hyper-perfusion. Further, these mechanisms are vital to the network behavior of the microvasculature within the various tissues. These local adjustments are acutely provided through changes in vessel diameter that occur via modulating the degree of smooth muscle contraction in the walls of resistance vessels. Initiating these changes in vascular tone are a variety of factors from within the local environment—both mechanical (pressure and shear stress) and chemical (metabolites and locally produced paracrine factors) in nature. While vascular smooth muscle cells of the resistance vessels supply the contractile element they are, in many situations, closely regulated by the endothelial cells.

Importantly, while local control mechanisms respond to acutely affect changes in tissue hemodynamics, they do not exist in isolation. Considerable interaction occurs with intrinsic regulatory mechanisms provided by neural and endocrine signaling. Further, the microvasculature itself demonstrates a plasticity over a longer time-frame by adaptive growth changes that alter the number of vessels available for flow and surface area for exchange. In pathophysiological states overt remodeling (including changes in wall thickness, lumen diameter or collateral formation) may occur to adapt to alterations in physical forces (for example in hypertension) and degrees of ischemia (as seen in vessel stenosis).

References

[1]   Handbook of Physiology. Microcirculation (2nd ed), edited by Tuma RF, Duran WN and Ley K. San Diego: Academic Press, 2008.

[2]   Handbook of Physiology. The Cardiovascular System, Vascular Smooth Muscle, American Physiological Society, Bethesda, MD. 1980.

[3]   Aalkjaer C, Heagerty AM, Petersen KK, Swales JD, and Mulvany MJ. Evidence for increased media thickness, increased neuronal amine uptake, and depressed excitation—contraction coupling in isolated resistance vessels from essential hypertensives. *Circ. Res.* 61: pp. 181–6, 1987.

[4]   Adebiyi A, Narayanan D, and Jaggar JH. Caveolin-1 assembles type 1 inositol 1,4,5-trisphosphate receptors and canonical transient receptor potential 3 channels into a functional signaling complex in arterial smooth muscle cells. *J. Biol. Chem.* 286: pp. 4341–8, 2011.

[5]   Adebiyi A, Zhao G, Narayanan D, Thomas-Gatewood CM, Bannister JP, and Jaggar JH. Isoform-selective physical coupling of TRPC3 channels to IP3 receptors in smooth muscle cells regulates arterial contractility. *Circ. Res.* 106: pp. 1603–12, 2010.

[6]   Albert AP, and Large WA. Signal transduction pathways and gating mechanisms of native TRP-like cation channels in vascular myocytes. *J. Physiol.* 570: pp. 45–51, 2006.

[7]   Alexander SP, Mathie A, and Peters JA. Guide to Receptors and Channels (GRAC), 5th edition. *Br. J. Pharmacol.* 164 Suppl 1: pp. S1–324, 2011.

[8]   Ando J, and Yamamoto K. Vascular mechanobiology: endothelial cell responses to fluid shear stress. *Circ. J.* 73: pp. 1983–92, 2009.

[9]   Antonios TF, Rattray FM, Singer DR, Markandu ND, Mortimer PS, and MacGregor GA. Rarefaction of skin capillaries in normotensive offspring of individuals with essential hypertension. *Heart* 89: pp. 175–8, 2003.

[10]   Armstrong ML, Dua AK, and Murrant CL. Potassium initiates vasodilatation induced by a single skeletal muscle contraction in hamster cremaster muscle. *J. Physiol.* 581: pp. 841–52, 2007.

[11]   Baez S. An open cremaster muscle preparation for the study of blood vessels by in vivo microscopy. *Microvasc. Res.* 5: pp. 384–94, 1973.

[12]   Baez S. Recording of microvascular dimensions with an image-splitter television micro-scope. *J. Appl. Physiol.* 21: pp. 299–301, 1966.

[13]   Bagher P, and Segal SS. Regulation of blood flow in the microcirculation: role of conducted vasodilation. *Acta. Physiol. (Oxf.)* 202: pp. 271–84, 2011.

[14]   Bagi Z, Frangos JA, Yeh JC, White CR, Kaley G, and Koller A. PECAM-1 mediates NO-dependent dilation of arterioles to high temporal gradients of shear stress. *Arterioscler Thromb. Vasc. Biol.* 25: pp. 1590–5, 2005.

[15]   Bakker EN, Buus CL, Spaan JA, Perree J, Ganga A, Rolf TM, Sorop O, Bramsen LH, Mulvany MJ, and Vanbavel E. Small artery remodeling depends on tissue-type transgluta-minase. *Circ. Res.* 96: pp. 119–26, 2005.

[16]   Bakker EN, Pistea A, and VanBavel E. Transglutaminases in vascular biology: relevance for vascular remodeling and atherosclerosis. *J. Vasc. Res.* 45: pp. 271–8, 2008.

[17]   Ballard ST, Hill MA, and Meininger GA. Effect of vasodilation and vasoconstriction on microvascular pressures in skeletal muscle. *Microcirc. Endothelium Lymphatics* 7: pp. 109–31, 1991.

[18]   Balligand JL, Feron O, and Dessy C. eNOS activation by physical forces: from short-term regulation of contraction to chronic remodeling of cardiovascular tissues. *Physiol. Rev.* 89: pp. 481–534, 2009.

[19]   Baumbach GL, and Heistad DD. Effects of sympathetic stimulation and changes in arterial pressure on segmental resistance of cerebral vessels in rabbits and cats. *Circ. Res.* 52: pp. 527–33, 1983.

[20]   Bayliss WM. On the local reactions of the arterial wall to changes of internal pressure. *J. Physiol.* 28: pp. 220–31, 1902.

[21]   Bayliss WM, and Starling EH. Observations on Venous Pressures and their Relationship to Capillary Pressures. *J. Physiol.* 16: pp. 159–318, 1894.

[22]   Beleznai TZ, Yarova PL, Yuill KH, and Dora KA. Smooth muscle ca(2+) -activated and voltage-gated k(+) channels modulate conducted dilation in rat isolated small mesenteric arteries. *Microcirculation* 18: pp. 487–500, 2011.

[23]   Bender SB, Berwick ZC, Laughlin MH, and Tune JD. Functional contribution of P2Y1 receptors to the control of coronary blood flow. *J. Appl. Physiol.* 111: pp. 1744–50, 2011.

[24]   Bergfeld GR, and Forrester T. Release of ATP from human erythrocytes in response to a brief period of hypoxia and hypercapnia. *Cardiovasc. Res.* 26: pp. 40–7, 1992.

[25]   Bergstrom S, and Samuelsson B. Isolation of prostaglandin E1 from human seminal plasma. Prostaglandins and related factors. 11. *J. Biol. Chem.* 237: pp. 3005–6, 1962.

[26]   Berne RM. Regulation of Coronary Blood Flow. *Physiol. Rev.* 44: pp. 1–29, 1964.

[27]   Berne RM, Blackmon JR, and Gardner TH. Hypoxemia and coronary blood flow. *J. Clin. Invest.* 36: pp. 1101–6, 1957.

[28]   Berridge MJ, Bootman MD, and Roderick HL. Calcium signalling: dynamics, homeostasis and remodelling. *Nat. Rev. Mol. Cell. Biol.* 4: pp. 517–29, 2003.

[29]   Bishop JJ, Nance PR, Popel AS, Intaglietta M, and Johnson PC. Diameter changes in skeletal muscle venules during arterial pressure reduction. *Am. J. Physiol. Heart. Circ. Physiol.* 279: pp. H47–57, 2000.

[30]   Bjornberg J, Grande PO, Maspers M, and Mellander S. Site of autoregulatory reactions in the vascular bed of cat skeletal muscle as determined with a new technique for segmental vascular resistance recordings. *Acta. Physiol. Scand.* 133: pp. 199–210, 1988.

[31]   Bloch EH. In vivo microscopic observations of the circulating blood in acute myocardial infarction. *Am. J. Med. Sci.* 229: pp. 280–94, 1955.

[32]   Bohlen HG. Na+-induced intestinal interstitial hyperosmolality and vascular responses during absorptive hyperemia. *Am. J. Physiol.* 242: pp. H785–9, 1982.

[33]   Bohlen HG, and Gore RW. Comparison of microvascular pressures and diameters in the innervated and denervated rat intestine. *Microvasc. Res.* 14: pp. 251–64, 1977.

[34]   Bolton TB, Gordienko DV, Povstyan OV, Harhun MI, and Pucovsky V. Smooth muscle cells and interstitial cells of blood vessels. *Cell Calcium* 35: pp. 643–57, 2004.

[35]   Bolz SS, and Pohl U. Highly effective non-viral gene transfer into vascular smooth muscle cells of cultured resistance arteries demonstrated by genetic inhibition of sphingosine-1-phosphate-induced vasoconstriction. *J. Vasc. Res.* 40: pp. 399–405, 2003.

[36]   Bolz SS, Vogel L, Sollinger D, Derwand R, Boer C, Pitson SM, Spiegel S, and Pohl U. Sphingosine kinase modulates microvascular tone and myogenic responses through activation of RhoA/Rho kinase. *Circulation* 108: pp. 342–7, 2003.

[37]   Bond RF, Blackard RF, and Taxis JA. Evidence against oxygen being the primary factor governing autoregulation. *Am. J. Physiol.* 216: pp. 788–93, 1969.

[38]   Borders JL, and Granger HJ. An optical doppler intravital velocimeter. *Microvasc. Res.* 27: pp. 117–27, 1984.

[39]   Borisova L, Wray S, Eisner DA, and Burdyga T. How structure, Ca signals, and cellular communications underlie function in precapillary arterioles. *Circ. Res.* 105: pp. 803–10, 2009.

[40]   Borst HG, McGregor M, Whittenberger JL, and Berglund E. Influence of pulmonary arterial and left atrial pressures on pulmonary vascular resistance. *Circ. Res.* 4: pp. 393–9, 1956.

[41]   Brenner BM, Troy JL, and Daugharty TM. The dynamics of glomerular ultrafiltration in the rat. *J. Clin. Invest.* 50: pp. 1776–80, 1971.

[42]   Brevetti G, Silvestro A, Schiano V, and Chiariello M. Endothelial dysfunction and cardiovascular risk prediction in peripheral arterial disease: additive value of flow-mediated dilation to ankle-brachial pressure index. *Circulation* 108: pp. 2093–8, 2003.

[43]   Brownlee M. Biochemistry and molecular cell biology of diabetic complications. *Nature* 414: pp. 813–20, 2001.

[44]   Brownlee M. The pathobiology of diabetic complications: a unifying mechanism. *Diabetes* 54: pp. 1615–25, 2005.

[45]   Bunting S, Gryglewski R, Moncada S, and Vane JR. Arterial walls generate from prosta-glandin endoperoxides a substance (prostaglandin X) which relaxes strips of mesenteric and coeliac ateries and inhibits platelet aggregation. *Prostaglandins* 12: pp. 897–913, 1976.

[46]   Burns WR, Cohen KD, and Jackson WF. K+-induced dilation of hamster cremasteric arte-rioles involves both the Na+/K+-ATPase and inward-rectifier K+ channels. *Microcirculation* 11: pp. 279–93, 2004.

[47]   Burrows ME, and Johnson PC. Diameter, wall tension, and flow in mesenteric arterioles during autoregulation. *Am. J. Physiol.* 241: pp. H829–37, 1981.

[48]   Calbet JA, and Joyner MJ. Disparity in regional and systemic circulatory capacities: do they affect the regulation of the circulation? *Acta. Physiol. (Oxf.)* 199: pp. 393–406, 2010.

[49]   Campochiaro PA. Ocular neovascularisation and excessive vascular permeability. *Expert Opin. Biol. Ther.* 4: pp. 1395–402, 2004.

[50]   Carr P, Graves JE, and Poston L. Carbon dioxide induced vasorelaxation in rat mesenteric small arteries precontracted with noradrenaline is endothelium dependent and mediated by nitric oxide. *Pflugers. Arch.* 423: pp. 343–5, 1993.

[51]   Case RB, and Greenberg H. The response of canine coronary vascular resistance to local alterations in coronary arterial P CO_2. *Circ. Res.* 39: pp. 558–66, 1976.

[52]   Casey DP, and Joyner MJ. Contribution of adenosine to compensatory dilation in hypo-perfused contracting human muscles is independent of nitric oxide. *J. Appl. Physiol.* 110: pp. 1181–9, 2011.

[53]   Casey DP, and Joyner MJ. Local control of skeletal muscle blood flow during exercise: influ-ence of available oxygen. *J. Appl. Physiol.* 111: pp. 1527–38, 2011.

[54]   Celermajer DS, Sorensen KE, Gooch VM, Spiegelhalter DJ, Miller OI, Sullivan ID, Lloyd JK, and Deanfield JE. Non-invasive detection of endothelial dysfunction in children and adults at risk of atherosclerosis. *Lancet* 340: pp. 1111–5, 1992.

[55]   Chachisvilis M, Zhang YL, and Frangos JA. G protein-coupled receptors sense fluid shear stress in endothelial cells. *Proc. Natl. Acad. Sci. U S A* 103: pp. 15463–8, 2006.

[56]   Chadha PS, Liu L, Rikard-Bell M, Senadheera S, Howitt L, Bertrand RL, Grayson TH, Murphy TV, and Sandow SL. Endothelium-dependent vasodilation in human mesenteric artery is primarily mediated by myoendothelial gap junctions intermediate conductance calcium-activated K+ channel and nitric oxide. *J. Pharmacol. Exp. Ther.* 336: pp. 701–8, 2011.

[57]   Chen, II, Prewitt RL, and Dowell RF. Microvascular rarefaction in spontaneously hyper-tensive rat cremaster muscle. *Am. J. Physiol.* 241: pp. H306–10, 1981.

[58]   Chen Q, and Anderson DR. Effect of CO_2 on intracellular pH and contraction of retinal capillary pericytes. *Invest. Ophthalmol. Vis. Sci.* 38: pp. 643–51, 1997.

[59]   Chen TT, Luykenaar KD, Walsh EJ, Walsh MP, and Cole WC. Key role of Kv1 channels in vasoregulation. *Circ. Res.* 99: pp. 53–60, 2006.

[60]   Chen YL, Wolin MS, and Messina EJ. Evidence for cGMP mediation of skeletal muscle arteriolar dilation to lactate. *J. Appl. Physiol.* 81: pp. 349–54, 1996.

[61]   Cheng H, Lederer WJ, and Cannell MB. Calcium sparks: elementary events underlying excitation-contraction coupling in heart muscle. *Science* 262: pp. 740–4, 1993.

[62]   Chi JT, Chang HY, Haraldsen G, Jahnsen FL, Troyanskaya OG, Chang DS, Wang Z, Rockson SG, van de Rijn M, Botstein D, and Brown PO. Endothelial cell diversity revealed by global expression profiling. *Proc. Natl. Acad. Sci. U S A* 100: pp. 10623–8, 2003.

[63]   Chilian WM, Eastham CL, and Marcus ML. Microvascular distribution of coronary vascular resistance in beating left ventricle. *Am. J. Physiol.* 251: pp. H779–88, 1986.

[64]   Chilton L, Loutzenhiser K, Morales E, Breaks J, Kargacin GJ, and Loutzenhiser R. Inward rectifier K(+) currents and Kir2.1 expression in renal afferent and efferent arterioles. *J. Am. Soc. Nephrol.* 19: pp. 69–76, 2008.

[65]   Christensen KL, and Mulvany MJ. Location of resistance arteries. *J. Vasc. Res.* 38: pp. 1–12, 2001.

[66]   Cipolla M.J. The Cerebral Circulation. *Colloquium Series on Integrated Systems Physiology: From Molecule to Function to Disease*, ed. D.N. Granger and J.P. Granger. Princeton, NJ: Morgan & Claypool Life Sciences, 2009. doi:10.4199/C0005ED1V01Y200912ISP002.

[67]   Clapham DE, Runnels LW, and Strubing C. The TRP ion channel family. *Nat. Rev. Neurosci.* 2: pp. 387–96, 2001.

[68]   Clark ER. The transparent chamber technique for the microscopic study of living blood vessels. *Anat. Rec.* 120: pp. 241–51, 1954.

[69]   Clifford PS. Local control of blood flow. *Adv. Physiol. Educ.* 35: pp. 5–15, 2011.

[70]   Clifford PS. Skeletal muscle vasodilatation at the onset of exercise. *J. Physiol.* 583: pp. 825–33, 2007.

[71]   Clifford PS, Ella SR, Stupica AJ, Nourian Z, Li M, Martinez-Lemus LA, Dora KA, Yang Y, Davis MJ, Pohl U, Meininger GA, and Hill MA. Spatial distribution and mechanical function of elastin in resistance arteries: a role in bearing longitudinal stress. *Arterioscler. Thromb. Vasc. Biol.* 31: pp. 2889–96, 2011.

[72]   Coen M, Gabbiani G, and Bochaton-Piallat ML. Myofibroblast-mediated adventitial remodeling: an underestimated player in arterial pathology. *Arterioscler. Thromb. Vasc. Biol.* 31: pp. 2391–6, 2011.

[73]   Cohen KD, and Jackson WF. Hypoxia inhibits contraction but not calcium channel

currents or changes in intracellular calcium in arteriolar muscle cells. *Microcirculation* 10: pp. 133–41, 2003.

[74]   Cole WC, and Welsh DG. Role of myosin light chain kinase and myosin light chain phosphatase in the resistance arterial myogenic response to intravascular pressure. *Arch. Biochem. Biophys.* 510: pp. 160–73, 2011.

[75]   Cox RH. Molecular determinants of voltage-gated potassium currents in vascular smooth muscle. *Cell. Biochem. Biophys.* 42: pp. 167–95, 2005.

[76]   Crecelius AR, Kirby BS, Richards JC, Garcia LJ, Voyles WF, Larson DG, Luckasen GJ, and Dinenno FA. Mechanisms of ATP-mediated vasodilation in humans: modest role for nitric oxide and vasodilating prostaglandins. *Am. J. Physiol. Heart. Circ. Physiol.* 301: pp. H1302–10, 2011.

[77]   D'Angelo G, Mogford JE, Davis GE, Davis MJ, and Meininger GA. Integrin-mediated reduction in vascular smooth muscle [Ca2+]i induced by RGD-containing peptide. *Am. J. Physiol.* 272: pp. H2065–70, 1997.

[78]   Daly SM, and Leahy MJ. 'Go with the flow': A review of methods and advancements in blood flow imaging. *J. Biophotonics*, 2012.

[79]   Daub B, and Ganitkevich V. An estimate of rapid cytoplasmic calcium buffering in a single smooth muscle cell. *Cell Calcium* 27: pp. 3–13, 2000.

[80]   Davies PF. Flow-mediated endothelial mechanotransduction. *Physiol. Rev.* 75: pp. 519–60, 1995.

[81]   Davis GE. Matricryptic sites control tissue injury responses in the cardiovascular system: relationships to pattern recognition receptor regulated events. *J. Mol. Cell. Cardiol.* 48: pp. 454–60, 2010.

[82]   Davis MJ. Control of bat wing capillary pressure and blood flow during reduced perfusion pressure. *Am. J. Physiol.* 255: pp. H1114–29, 1988.

[83]   Davis MJ. Microvascular control of capillary pressure during increases in local arterial and venous pressure. *Am. J. Physiol.* 254: pp. H772–84, 1988.

[84]   Davis MJ. Myogenic response gradient in an arteriolar network. *Am. J. Physiol.* 264: pp. H2168–79, 1993.

[85]   Davis MJ. Perspective: physiological role(s) of the vascular myogenic response. *Microcirculation* 19: pp. 99–114, 2012.

[86]   Davis MJ, Donovitz JA, and Hood JD. Stretch-activated single-channel and whole cell currents in vascular smooth muscle cells. *Am. J. Physiol.* 262: pp. C1083–8, 1992.

[87]   Davis MJ, and Hill MA. Signaling mechanisms underlying the vascular myogenic response. *Physiol. Rev.* 79: pp. 387–423, 1999.

[88]   Davis MJ, Hill MA, and Kuo L. Local regulation of microvascular perfusion. In: *Handbook*

of Physiology. Microcirculation (2nd ed.), edited by Tuma RF, Duran WN and Ley K. San Diego: Academic Press, 2008.

[89]   Davis MJ, Joyner WL, and Gilmore JP. Microvascular pressure distribution and responses of pulmonary allografts and cheek pouch arterioles in the hamster to oxygen. *Circ. Res.* 49: pp. 125–32, 1981.

[90]   Davis MJ, and Meininger GA. Myogenic response in microvascular networks. In: *Mechano-reception by the Vascular Wall*, edited by Rubanyi GM. Mount Kisco, NY: Futura Publishing Company, 1993, pp. 37–60.

[91]   Davis MJ, and Sikes PJ. A rate-sensitive component to the myogenic response is absent from bat wing arterioles. *Am. J. Physiol.* 256: pp. H32–40, 1989.

[92]   Dawes GS. The vaso-dilator action of potassium. *J. Physiol.* 99: pp. 224–38, 1941.

[93]   de Wit C. Different pathways with distinct properties conduct dilations in the microcirculation in vivo. *Cardiovasc. Res.* 85: pp. 604–13, 2010.

[94]   de Wit C, and Griffith TM. Connexins and gap junctions in the EDHF phenomenon and conducted vasomotor responses. *Pflugers. Arch.* 459: pp. 897–914, 2010.

[95]   de Wit C, and Wolfle SE. EDHF and gap junctions: important regulators of vascular tone within the microcirculation. *Curr. Pharm. Biotechnol.* 8: pp. 11–25, 2007.

[96]   DeFily DV, and Chilian WM. Coronary microcirculation: autoregulation and metabolic control. *Basic. Res. Cardiol.* 90: pp. 112–8, 1995.

[97]   DeLano FA, Schmid-Schonbein GW, Skalak TC, and Zweifach BW. Penetration of the systemic blood pressure into the microvasculature of rat skeletal muscle. *Microvasc. Res.* 41: pp. 92–110, 1991.

[98]   Di Wang H, Ratsep MT, Chapman A, and Boyd R. Adventitial fibroblasts in vascular structure and function: the role of oxidative stress and beyond. *Can. J. Physiol. Pharmacol.* 88: pp. 177–86, 2010.

[99]   Domeier TL, and Segal SS. Electromechanical and pharmacomechanical signalling pathways for conducted vasodilatation along endothelium of hamster feed arteries. *J. Physiol.* 579: pp. 175–86, 2007.

[100] Dora KA, Doyle MP, and Duling BR. Elevation of intracellular calcium in smooth muscle causes endothelial cell generation of NO in arterioles. *Proc. Natl. Acad. Sci. U S A* 94: pp. 6529–34, 1997.

[101] Dora KA, Sandow SL, Gallagher NT, Takano H, Rummery NM, Hill CE, and Garland CJ. Myoendothelial gap junctions may provide the pathway for EDHF in mouse mesenteric artery. *J. Vasc. Res.* 40: pp. 480–90, 2003.

[102] Dore-Duffy P, and Cleary K. Morphology and properties of pericytes. *Methods Mol. Biol.* 686: pp. 49–68, 2011.

[103] Doyle MP, Linden J, and Duling BR. Nucleoside-induced arteriolar constriction: a mast cell-dependent response. *Am. J. Physiol.* 266: pp. H2042–50, 1994.

[104] Drummond HA, Grifoni SC, and Jernigan NL. A new trick for an old dogma: ENaC proteins as mechanotransducers in vascular smooth muscle. *Physiology (Bethesda)* 23: pp. 23–31, 2008.

[105] Duff F, Patterson GC, and Shepherd JT. A quantitative study of the response to adenosine triphosphate of the blood vessels of the human hand and forearm. *J. Physiol.* 125: 581–9, 1954.

[106] Duling BR, Gore RW, Dacey RG, Jr., and Damon DN. Methods for isolation, cannulation, and in vitro study of single microvessels. *Am. J. Physiol.* 241: pp. H108–16, 1981.

[107] Duling BR, and Staples E. Microvascular effects of hypertonic solutions in the hamster. *Microvasc. Res.* 11: pp. 51–6, 1976.

[108] Dull RO, Mecham I, and McJames S. Heparan sulfates mediate pressure-induced increase in lung endothelial hydraulic conductivity via nitric oxide/reactive oxygen species. *Am. J. Physiol. Lung. Cell Mol. Physiol.* 292: pp. L1452–8, 2007.

[109] Duncker DJ, Laxson DD, Lindstrom P, and Bache RJ. Endogenous adenosine and coronary vasoconstriction in hypoperfused myocardium during exercise. *Cardiovasc. Res.* 27: pp. 1592–7, 1993.

[110] Durand S, Zhang R, Cui J, Wilson TE, and Crandall CG. Evidence of a myogenic response in vasomotor control of forearm and palm cutaneous microcirculations. *J. Appl. Physiol.* 97: pp. 535–9, 2004.

[111] Earley S, and Brayden JE. Transient receptor potential channels and vascular function. *Clin. Sci. (Lond.)* 119: pp. 19–36, 2010.

[112] Earley S, Straub SV, and Brayden J. Protein Kinase C Regulates Vascular Myogenic Tone Through Activation of TRPM4. *Am. J. Physiol. Heart Circ. Physiol.*, 2007.

[113] Earley S, Waldron BJ, and Brayden JE. Critical role for transient receptor potential channel TRPM4 in myogenic constriction of cerebral arteries. *Circ. Res.* 95: pp. 922–9, 2004.

[114] Edwards G, Feletou M, and Weston AH. Endothelium-derived hyperpolarising factors and associated pathways: a synopsis. *Pflugers. Arch.* 459: pp. 863–79, 2010.

[115] Edwards RM. Segmental effects of norepinephrine and angiotensin II on isolated renal microvessels. *Am. J. Physiol.* 244: pp. F526–34, 1983.

[116] El-Yazbi AF, Johnson RP, Walsh EJ, Takeya K, Walsh MP, and Cole WC. Pressure-dependent contribution of Rho kinase-mediated calcium sensitization in serotonin-evoked vasoconstriction of rat cerebral arteries. *J. Physiol.* 588: pp. 1747–62, 2010.

[117] Ella SR, Davis MJ, Meininger GA, Yang Y, Dora KA, and Hill MA. Mechanisms underlying smooth muscle Ca2+ waves in cremaster muscle arterioles. *Faseb. J.*, 2009.

[118] Ellsworth ML, Forrester T, Ellis CG, and Dietrich HH. The erythrocyte as a regulator of vascular tone. *Am. J. Physiol.* 269: pp. H2155–61, 1995.

[119] Erickson HP. Stretching fibronectin. *J. Muscle Res. Cell. Motil.* 23: pp. 575–80, 2002.

[120] Erlinge D, and Burnstock G. P2 receptors in cardiovascular regulation and disease. *Puriner-gic Signal* 4: pp. 1–20, 2008.

[121] Evans RJ, and Surprenant A. Vasoconstriction of guinea-pig submucosal arterioles follow-ing sympathetic nerve stimulation is mediated by the release of ATP. *Br. J. Pharmacol.* 106: pp. 242–9, 1992.

[122] Fagrell B. Microcirculatory methods for the clinical assessment of hypertension, hypoten-sion, and ischemia. *Ann. Biomed. Eng.* 14: pp. 163–73, 1986.

[123] Falcone JC, Davis MJ, and Meininger GA. Endothelial independence of myogenic re-sponse in isolated skeletal muscle arterioles. *Am. J. Physiol.* 260: pp. H130–5, 1991.

[124] Falcone JC, Granger HJ, and Meininger GA. Enhanced myogenic activation in skeletal muscle arterioles from spontaneously hypertensive rats. *Am. J. Physiol.* 265: pp. H1847–55, 1993.

[125] Falloon BJ, and Heagerty AM. In vitro perfusion studies of human resistance artery func-tion in essential hypertension. *Hypertension* 24: pp. 16–23, 1994.

[126] Falloon BJ, Stephens N, Tulip JR, and Heagerty AM. Comparison of small artery sensi-tivity and morphology in pressurized and wire-mounted preparations. *Am. J. Physiol.* 268: pp. H670–8, 1995.

[127] Falsini B, Anselmi GM, Marangoni D, D'Esposito F, Fadda A, Di Renzo A, Campos EC, and Riva CE. Subfoveal choroidal blood flow and central retinal function in retinitis pig-mentosa. *Invest. Ophthalmol. Vis. Sci.* 52: pp. 1064–9, 2011.

[128] Farkas E, and Luiten PG. Cerebral microvascular pathology in aging and Alzheimer's dis-ease. *Prog. Neurobiol.* 64: pp. 575–611, 2001.

[129] Feletou M. The Endothelium, Part I: Multiple Functions of the Endothelial Cells—Fo-cus on Endothelium-Derived Vasoactive Mediators. *Colloquium Series on Integrated Systems Physiology: From Molecule to Function to Disease*, ed. D.N. Granger and J.P. Granger. Prince-ton, NJ: Morgan & Claypool Life Sciences, 2011. doi:10.4199/C00031ED1V01Y201105 ISP019.

[130] Feletou M. Calcium-activated potassium channels and endothelial dysfunction: therapeutic options? *Br. J. Pharmacol.* 156: pp. 545–62, 2009.

[131] Feletou M, Huang Y, and Vanhoutte PM. Endothelium-mediated control of vascular tone: COX-1 and COX-2 products. *Br. J. Pharmacol.*, 2011.

[132] Feletou M, and Vanhoutte PM. EDHF: an update. *Clin. Sci. (Lond.)* 117: pp. 139–55, 2009.

[133] Fog M. Cerebral circulation: II. Reaction of pial arteries to increase in blood pressure. *AMA Arch. Neurol. Psychiat.* 41: p. 260, 1939.

[134] Folkow B. Intravascular pressure as a factor regulating the tone of the small vessels. *Acta. Physiol. Scand.* 17: pp. 289–310, 1949.

[135] Folkow B. "Structural factor" in primary and secondary hypertension. *Hypertension* 16: pp. 89–101, 1990.

[136] Folkow B. A study of the factors influencing the tone of denervated blood vessels perfused at various pressures. *Acta. Physiol. Scand.* 27: pp. 99–117, 1952.

[137] Fox JR, and Wiederhielm CA. Characteristics of the servo-controlled micropipet pressure system. *Microvasc. Res.* 5: pp. 324–35, 1973.

[138] Franco-Obregon A, and Lopez-Barneo J. Low PO2 inhibits calcium channel activity in arterial smooth muscle cells. *Am. J. Physiol.* 271: pp. H2290–9, 1996.

[139] Fronek K, and Zweifach BW. Microvascular pressure distribution in skeletal muscle and the effect of vasodilation. *Am. J. Physiol.* 228: pp. 791–6, 1975.

[140] Furchgott RF, and Zawadzki JV. The obligatory role of endothelial cells in the relaxation of arterial smooth muscle by acetylcholine. *Nature* 288: pp. 373–6, 1980.

[141] Garland CJ, Hiley CR, and Dora KA. EDHF: spreading the influence of the endothelium. *Br. J. Pharmacol.* 164: pp. 839–52, 2011.

[142] Gaskell WH. On the Tonicity of the Heart and Blood Vessels. *J. Physiol.* 3: pp. 48–92 16, 1880.

[143] Gebremedhin D, Lange AR, Lowry TF, Taheri MR, Birks EK, Hudetz AG, Narayanan J, Falck JR, Okamoto H, Roman RJ, Nithipatikom K, Campbell WB, and Harder DR. Production of 20-HETE and its role in autoregulation of cerebral blood flow. *Circ. Res.* 87: pp. 60–65, 2000.

[144] Gimbrone MA, Jr., and Alexander RW. Angiotensin II stimulation of prostaglandin production in cultured human vascular endothelium. *Science* 189: pp. 219–20, 1975.

[145] Glagov S, Vito R, Giddens DP, and Zarins CK. Micro-architecture and composition of artery walls: relationship to location, diameter and the distribution of mechanical stress. *J. Hypertens. Suppl.* 10: pp. S101–4, 1992.

[146] Gollasch M. Vasodilator signals from perivascular adipose tissue. *Br. J. Pharmacol.*, 2011.

[147] Goodman AH, Guyton AC, Drake R, and Loflin JH. A television method for measuring capillary red cell velocities. *J. Appl. Physiol.* 37: pp. 126–30, 1974.

[148] Gore RW. Pressures in cat mesenteric arterioles and capillaries during changes in systemic arterial blood pressure. *Circ. Res.* 34: pp. 581–91, 1974.

[149] Gore RW, and Bohlen HG. Pressure regulation in the microcirculation. *Fed. Proc.* 34: pp. 2031–7, 1975.

[150] Gorenne I, Su X, and Moreland RS. Caldesmon phosphorylation is catalyzed by two kinases in permeabilized and intact vascular smooth muscle. *J. Cell. Physiol.* 198: pp. 461–9, 2004.

[151] Granger HJ, and Guyton AC. Autoregulation of the total systemic circulation following destruction of the central nervous system in the dog. *Circ. Res.* 25: pp. 379–88, 1969.

[152] Grayson TH, Haddock RE, Murray TP, Wojcikiewicz RJ, and Hill CE. Inositol 1,4,5-trisphosphate receptor subtypes are differentially distributed between smooth muscle and endothelial layers of rat arteries. *Cell Calcium* 36: pp. 447–58, 2004.

[153] Green DJ, Jones H, Thijssen D, Cable NT, and Atkinson G. Flow-mediated dilation and cardiovascular event prediction: does nitric oxide matter? *Hypertension* 57: pp. 363–9, 2011.

[154] Greensmith JE, and Duling BR. Morphology of the constricted arteriolar wall: physiological implications. *Am. J. Physiol.* 247: pp. H687–98, 1984.

[155] Greenstein AS, Khavandi K, Withers SB, Sonoyama K, Clancy O, Jeziorska M, Laing I, Yates AP, Pemberton PW, Malik RA, and Heagerty AM. Local inflammation and hypoxia abolish the protective anticontractile properties of perivascular fat in obese patients. *Circulation* 119: pp. 1661–70, 2009.

[156] Greis C. Quantitative evaluation of microvascular blood flow by contrast-enhanced ultrasound (CEUS). *Clin. Hemorheol. Microcirc.* 49: pp. 137–49, 2011.

[157] Gudi S, Nolan JP, and Frangos JA. Modulation of GTPase activity of G proteins by fluid shear stress and phospholipid composition. *Proc. Natl. Acad. Sci. U S A* 95: pp. 2515–9, 1998.

[158] Gudi SR, Clark CB, and Frangos JA. Fluid flow rapidly activates G proteins in human endothelial cells. Involvement of G proteins in mechanochemical signal transduction. *Circ. Res.* 79: pp. 834–9, 1996.

[159] Gunst SJ, and Zhang W. Actin cytoskeletal dynamics in smooth muscle: a new paradigm for the regulation of smooth muscle contraction. *Am. J. Physiol. Cell Physiol* 295: pp. C576–87, 2008.

[160] Gurevicius J, Salem MR, Metwally AA, Silver JM, and Crystal GJ. Contribution of nitric oxide to coronary vasodilation during hypercapnic acidosis. *Am. J. Physiol.* 268: pp. H39–47, 1995.

[161] Gutman GA, Chandy KG, Grissmer S, Lazdunski M, McKinnon D, Pardo LA, Robertson GA, Rudy B, Sanguinetti MC, Stuhmer W, and Wang X. International Union of Pharmacology. LIII. Nomenclature and molecular relationships of voltage-gated potassium channels. *Pharmacol. Rev.* 57: pp. 473–508, 2005.

[162] Haddock RE, and Hill CE. Rhythmicity in arterial smooth muscle. *J. Physiol.* 566: pp. 645–56, 2005.

[163] Hall JE. Regulation of renal hemodynamics. *Int. Rev. Physiol.* 26: pp. 243–321, 1982.

[164] Halpern W, and Osol G. Influence of transmural pressure of myogenic responses of isolated cerebral arteries of the rat. *Ann. Biomed. Eng.* 13: pp. 287–93, 1985.

[165] Harder DR. Pressure-dependent membrane depolarization in cat middle cerebral artery. *Circ. Res.* 55: pp. 197–202, 1984.

[166] Harder DR, Narayanan J, Birks EK, Liard JF, Imig JD, Lombard JH, Lange AR, and Roman RJ. Identification of a putative microvascular oxygen sensor. *Circ. Res.* 79: pp. 54–61, 1996.

[167] Harder DR, Narayanan J, and Gebremedhin D. Pressure-induced myogenic tone and role of 20-HETE in mediating autoregulation of cerebral blood flow. *Am. J. Physiol. Heart Circ. Physiol.* 300: pp. H1557–65, 2011.

[168] Harhun MI, Gordienko DV, Povstyan OV, Moss RF, and Bolton TB. Function of interstitial cells of Cajal in the rabbit portal vein. *Circ. Res.* 95: pp. 619–26, 2004.

[169] Harhun MI, Szewczyk K, Laux H, Prestwich SA, Gordienko DV, Moss RF, and Bolton TB. Interstitial cells from rat middle cerebral artery belong to smooth muscle cell type. *J. Cell. Mol. Med.* 13: pp. 4532–9, 2009.

[170] Hartmann H, Orskov SL, and Rein H. Die Gefassreaktionen der Niere im Verlaufe allgemeiner Kreislauf-Regulationsvorgange. *Pfluegers. Arch.* 238: pp. 239–50, 1937.

[171] Hartmannsgruber V, Heyken WT, Kacik M, Kaistha A, Grgic I, Harteneck C, Liedtke W, Hoyer J, and Kohler R. Arterial response to shear stress critically depends on endothelial TRPV4 expression. *PLoS One* 2: p. e827, 2007.

[172] Hassan AA, and Tooke JE. Mechanism of the postural vasoconstrictor response in the human foot. *Clin. Sci. (Lond.)* 75: pp. 379–87, 1988.

[173] Hein TW, Belardinelli L, and Kuo L. Adenosine A(2A) receptors mediate coronary microvascular dilation to adenosine: role of nitric oxide and ATP-sensitive potassium channels. *J. Pharmacol. Exp. Ther.* 291: pp. 655–64, 1999.

[174] Hein TW, Xu W, and Kuo L. Dilation of retinal arterioles in response to lactate: role of nitric oxide, guanylyl cyclase, and ATP-sensitive potassium channels. *Invest. Ophthalmol. Vis. Sci.* 47: pp. 693–9, 2006.

[175] Hill MA, and Davis MJ. Coupling a change in intraluminal pressure to vascular smooth muscle depolarization: still stretching for an explanation. *Am. J. Physiol. Heart Circ. Physiol.* 292: pp. H2570–2, 2007.

[176] Hill MA, Davis MJ, Meininger GA, Potocnik SJ, and Murphy TV. Arteriolar myogenic signalling mechanisms: Implications for local vascular function. *Clin. Hemorheol. Microcirc.* 34: pp. 67–79, 2006.

[177] Hill MA, Falcone JC, and Meininger GA. Evidence for protein kinase C involvement in arteriolar myogenic reactivity. *Am. J. Physiol.* 259: pp. H1586–94, 1990.

[178] Hill MA, and Meininger GA. Arteriolar vascular smooth muscle cells: Mechanotransducers in a complex environment. *Int. J. Biochem. Cell. Biol.* 44: pp. 1505–10, 2012.

[179] Hill MA, and Meininger GA. Calcium entry and myogenic phenomena in skeletal muscle arterioles. *Am. J. Physiol.* 267: pp. H1085–92, 1994.

[180] Hill MA, Meininger GA, Davis MJ, and Laher I. Therapeutic potential of pharmacologically targeting arteriolar myogenic tone. *Trends Pharmacol. Sci.* 30: pp. 363–74, 2009.

[181] Hill MA, Sun Z, Martinez-Lemus L, and Meininger GA. New technologies for dissecting the arteriolar myogenic response. *Trends Pharmacol. Sci.*, 2007.

[182] Hill MA, Zou H, Davis MJ, Potocnik SJ, and Price S. Transient increases in diameter and [Ca(2+)](i) are not obligatory for myogenic constriction. *Am. J. Physiol. Heart Circ. Physiol.* 278: pp. H345–52, 2000.

[183] Hilton SM. On the increase in muscle blood flow following contraction. *J. Physiol.* 117: pp. 13p–4p, 1952.

[184] Hisadome K, Koyama T, Kimura C, Droogmans G, Ito Y, and Oike M. Volume-regulated anion channels serve as an auto/paracrine nucleotide release pathway in aortic endothelial cells. *J. Gen. Physiol.* 119: pp. 511–20, 2002.

[185] Hnik P, Kriz N, Vyskocil F, Smiesko V, Mejsnar J, Ujec E, and Holas M. Work-induced potassium changes in muscle venous effluent blood measured by ion-specific electrodes. *Pflugers. Arch.* 338: pp. 177–81, 1973.

[186] Hoepfl B, Rodenwaldt B, Pohl U, and De Wit C. EDHF, but not NO or prostaglandins, is critical to evoke a conducted dilation upon ACh in hamster arterioles. *Am. J. Physiol. Heart Circ. Physiol.* 283: pp. H996–1004, 2002.

[187] Hoger JH, Ilyin VI, Forsyth S, and Hoger A. Shear stress regulates the endothelial Kir2.1 ion channel. *Proc. Natl. Acad. Sci. U S A* 99: pp. 7780–5, 2002.

[188] Horiuchi T, Dietrich HH, Hongo K, and Dacey RG, Jr. Mechanism of extracellular K+-induced local and conducted responses in cerebral penetrating arterioles. *Stroke* 33: pp. 2692–9, 2002.

[189] House SD, and Johnson PC. Diameter and blood flow of skeletal muscle venules during local flow regulation. *Am. J. Physiol.* 250: pp. H828–37, 1986.

[190] Hsu P, Albuquerque ML, and Leffler CW. Mechanisms of hypercapnia-stimulated PG production in piglet cerebral microvascular endothelial cells. *Am. J. Physiol.* 268: pp. H591–603, 1995.

[191] Hwa C, and Aird WC. The history of the capillary wall: doctors, discoveries, and debates. *Am. J. Physiol. Heart Circ. Physiol.* 293: pp. H2667–79, 2007.

[192] Ido Y, Chang K, Woolsey TA, and Williamson JR. NADH: sensor of blood flow need in brain, muscle, and other tissues. *FASEB. J.* 15: pp. 1419–21, 2001.

[193] Ijzerman RG, Serne EH, van Weissenbruch MM, de Jongh RT, and Stehouwer CD. Cigarette smoking is associated with an acute impairment of microvascular function in humans. *Clin. Sci. (Lond.)* 104: pp. 247–52, 2003.

[194] Intaglietta M, Pawula RF, and Tompkins WR. Pressure measurements in the mammalian microvasculature. *Microvasc. Res.* 2: pp. 212–20, 1970.

[195] Intaglietta M, and Tompkins WR. Micropressure measurement with 1 micron and smaller cannulae. *Microvasc. Res.* 3: pp. 211–4, 1971.

[196] Intaglietta M, and Tompkins WR. Microvascular measurements by video image shearing and splitting. *Microvasc. Res.* 5: pp. 309–12, 1973.

[197] Intaglietta M, and Tompkins WR. On-line measurement of microvascular dimensions by television microscopy. *J. Appl. Physiol.* 32: pp. 546–51, 1972.

[198] Ishiguro M, and Wellman GC. Cellular basis of vasospasm: role of small diameter arteries and voltage-dependent Ca2+ channels. *Acta. Neurochir. Suppl.* 104: pp. 95–8, 2008.

[199] Ishizaka H, and Kuo L. Endothelial ATP-sensitive potassium channels mediate coronary microvascular dilation to hyperosmolarity. *Am. J. Physiol.* 273: pp. H104–12, 1997.

[200] Jaap AJ, Hammersley MS, Shore AC, and Tooke JE. Reduced microvascular hyperaemia in subjects at risk of developing type 2 (non-insulin-dependent) diabetes mellitus. *Diabetologia* 37: pp. 214–6, 1994.

[201] Jackson WF. Potassium channels in the peripheral microcirculation. *Microcirculation* 12: pp. 113–27, 2005.

[202] Jain RK. Transport of molecules across tumor vasculature. *Cancer Metastasis Rev.* 6: pp. 559–93, 1987.

[203] Jarhult J, and Mellander S. Autoregulation of capillary hydrostatic pressure in skeletal muscle during regional arterial hypo- and hypertension. *Acta. Physiol. Scand.* 91: pp. 32–41, 1974.

[204] Johnson PC. Autoregulation of blood flow. *Circ. Res.* 59: pp. 483–95, 1986.

[205] Johnson PC. The Myogenic Response. In: *Handbook of Physiology, The Cardiovascular System, Vascular Smooth Muscle*, 1980, pp. 409–42.

[206] Johnson PC. Principles of peripheral circulatory control. In: *Peripheral Circulation*, edited by Johnson PC. New York: Wiley, 1978, pp. 111–39.

[207] Johnson PC. Review of Previous Studies and Current Theories of Autoregulation. *Circ. Res.* 15 Suppl: pp. 2–9, 1964.

[208] Johnson PC, Burton KS, Henrich H, and Henrich U. Effect of occlusion duration on reactive hyperemia in sartorius muscle capillaries. *Am. J. Physiol.* 230: pp. 715–9, 1976.

[209] Johnson PC, Waugh WH, and Hinshaw LB. Autoregulation of Blood Flow. *Science* 140: pp. 203–7, 1963.

[210] Johnson PC, and Wayland H. Regulation of blood flow in single capillaries. *Am. J. Physiol.* 212: pp. 1405–15, 1967.

[211] Johnson RP, El-Yazbi AF, Takeya K, Walsh EJ, Walsh MP, and Cole WC. Ca2+ sensitization owing to Rho kinase-dependent phosphorylation of MYPT1-T855 contributes to myogenic control of arterial diameter. *J. Physiol.*, 2009.

[212] Jones RD, and Berne RM. Evidence for a metabolic mechanism in autoregulation of blood flow in skeletal muscle. *Circ. Res.* 17: pp. 540–54, 1965.

[213] Kauffenstein G, Laher I, Matroughi K, Guerineau NC, and Henrion D. Emerging role of G protein-coupled receptors in microvascular myogenic tone. *Cardiovascular Research*, 2012.

[214] Kaufmann PA, Gnecchi-Ruscone T, di Terlizzi M, Schafers KP, Luscher TF, and Camici PG. Coronary heart disease in smokers: vitamin C restores coronary microcirculatory function. *Circulation* 102: pp. 1233–8, 2000.

[215] Khalil, R.A. Regulation of Vascular Smooth Muscle Function. *Colloquium Series on Integrated Systems Physiology: From Molecule to Function to Disease*, ed. D.N. Granger and J.P. Granger. Princeton, NJ: Morgan & Claypool Life Sciences, 2010. doi:10.4199/C00012 ED1V01Y201005ISP007.

[216] Khoobehi B, Peyman GA, and Vo K. Laser-triggered repetitive fluorescein angiography. *Ophthalmology* 99: pp. 72–9, 1992.

[217] Kim C, Ye F, and Ginsberg MH. Regulation of Integrin Activation. *Annu. Rev. Cell. Dev. Biol.*, 2011.

[218] Kim HR, Appel S, Vetterkind S, Gangopadhyay SS, and Morgan KG. Smooth muscle signalling pathways in health and disease. *J. Cell. Mol. Med.* 12: pp. 2165–80, 2008.

[219] Kimura G, and Brenner BM. Indirect assessment of glomerular capillary pressure from pressure-natriuresis relationship: comparison with direct measurements reported in rats. *Hypertens. Res.* 20: pp. 143–8, 1997.

[220] Kirchheim HR, Ehmke H, Hackenthal E, Lowe W, and Persson P. Autoregulation of renal blood flow, glomerular filtration rate and renin release in conscious dogs. *Pflugers. Arch.* 410: pp. 441–9, 1987.

[221] Knot HJ, and Nelson MT. Regulation of arterial diameter and wall [Ca2+] in cerebral arteries of rat by membrane potential and intravascular pressure. *J. Physiol.* 508 (Pt 1): pp. 199–209, 1998.

[222] Knot HJ, and Nelson MT. Regulation of membrane potential and diameter by voltage-dependent K+ channels in rabbit myogenic cerebral arteries. *Am. J. Physiol.* 269: pp. H348–55, 1995.

[223] Koch AR. Some Mathematical Forms of Autoregulatory Models. *Circ. Res.* 15 Suppl: pp. 269–78, 1964.

[224] Kohler R, Heyken WT, Heinau P, Schubert R, Si H, Kacik M, Busch C, Grgic I, Maier T, and Hoyer J. Evidence for a functional role of endothelial transient receptor potential V4 in shear stress-induced vasodilatation. *Arterioscler. Thromb. Vasc. Biol.* 26: pp. 1495–1502, 2006.

[225] Kohler R, and Ruth P. Endothelial dysfunction and blood pressure alterations in K+-channel transgenic mice. *Pflugers. Arch.* 459: pp. 969–76, 2010.

[226] Koller A, and Bagi Z. Nitric oxide and H_2O_2 contribute to reactive dilation of isolated coronary arterioles. *Am. J. Physiol. Heart. Circ. Physiol.* 287: pp. H2461–7, 2004.

[227] Koller A, and Bagi Z. On the role of mechanosensitive mechanisms eliciting reactive hyperemia. *Am. J. Physiol. Heart. Circ. Physiol.* 283: pp. H2250–9, 2002.

[228]   Koller A, and Kaley G. Endothelium regulates skeletal muscle microcirculation by a blood flow velocity-sensing mechanism. *Am. J. Physiol.* 258: pp. H916–20, 1990.

[229]   Koller A, Sun D, and Kaley G. Role of shear stress and endothelial prostaglandins in flow- and viscosity-induced dilation of arterioles in vitro. *Circ. Res.* 72: pp. 1276–84, 1993.

[230]   Kontos HA, Mauck HP, Jr., and Patterson JL, Jr. Mechanism of reactive hyperemia in limbs of anesthetized dogs. *Am. J. Physiol.* 209: pp. 1106–14, 1965.

[231]   Kontos HA, Wei EP, Raper AJ, Rosenblum WI, Navari RM, and Patterson JL, Jr. Role of tissue hypoxia in local regulation of cerebral microcirculation. *Am. J. Physiol.* 234: pp. H582–91, 1978.

[232]   Kotecha N, and Hill MA. Myogenic contraction in rat skeletal muscle arterioles: smooth muscle membrane potential and Ca(2+) signaling. *Am. J. Physiol. Heart. Circ. Physiol.* 289: pp. H1326–34, 2005.

[233]   Krogh A. Studies on the physiology of capillaries: II. The reactions to local stimuli of the blood-vessels in the skin and web of the frog. *J. Physiol.* 55: pp. 412–22, 1921.

[234]   Kuchan MJ, and Frangos JA. Role of calcium and calmodulin in flow-induced nitric oxide production in endothelial cells. *Am. J. Physiol.* 266: pp. C628–36, 1994.

[235]   Kuo IY, Wolfle SE, and Hill CE. T-type calcium channels and vascular function: the new kid on the block? *J. Physiol.* 589: pp. 783–95, 2011.

[236]   Kuo L, Chilian WM, and Davis MJ. Coronary arteriolar myogenic response is independent of endothelium. *Circ. Res.* 66: pp. 860–6, 1990.

[237]   Kuo L, Chilian WM, and Davis MJ. Interaction of pressure- and flow-induced responses in porcine coronary resistance vessels. *Am. J. Physiol.* 261: pp. H1706–15, 1991.

[238]   Kuo L, Davis MJ, and Chilian WM. Endothelial modulation of arteriolar tone. *News Physiol. Sci.* 7: pp. 5–9, 1992.

[239]   Kuo L, Davis MJ, and Chilian WM. Longitudinal gradients for endothelium-dependent and -independent vascular responses in the coronary microcirculation. *Circulation* 92: pp. 518–25, 1995.

[240]   Kusano Y, Echeverry G, Miekisiak G, Kulik TB, Aronhime SN, Chen JF, and Winn HR. Role of adenosine A2 receptors in regulation of cerebral blood flow during induced hypo-tension. *J. Cereb. Blood. Flow. Metab.* 30: pp. 808–15, 2010.

[241]   Laird JD, Breuls PN, van der Meer P, and Spaan JA. Can a single vasodilator be respon-sible for both coronary autoregulation and metabolic vasodilation? *Basic Res. Cardiol.* 76: pp. 354–8, 1981.

[242]   Lange A, Gebremedhin D, Narayanan J, and Harder D. 20-Hydroxyeicosatetraenoic acid-induced vasoconstriction and inhibition of potassium current in cerebral vascular smooth muscle is dependent on activation of protein kinase C. *J. Biol. Chem.* 272: pp. 27345–52, 1997.

[243] Langille BL, and O'Donnell F. Reductions in arterial diameter produced by chronic decreases in blood flow are endothelium-dependent. *Science* 231: pp. 405–7, 1986.

[244] Laughlin MH, Davis MJ, Secher NH, van Lieshout JJ, Simmons GH, Bender SB, Padilla J, Bache RJ, Merkus D, and Duncker DJ. Peripheral circulation. In: *Comprehensive Physiology*, 2012.

[245] Laxson DD, Homans DC, and Bache RJ. Inhibition of adenosine-mediated coronary vasodilation exacerbates myocardial ischemia during exercise. *Am. J. Physiol.* 265: pp. H1471–7, 1993.

[246] Ledoux J, Taylor MS, Bonev AD, Hannah RM, Solodushko V, Shui B, Tallini Y, Kotlikoff MI, and Nelson MT. Functional architecture of inositol 1,4,5-trisphosphate signaling in restricted spaces of myoendothelial projections. *Proc. Natl. Acad. Sci. U S A* 105: pp. 9627–32, 2008.

[247] Ledoux J, Werner ME, Brayden JE, and Nelson MT. Calcium-activated potassium channels and the regulation of vascular tone. *Physiology (Bethesda)* 21: pp. 69–78, 2006.

[248] Lee YC, Chang HH, Chiang CL, Liu CH, Yeh JI, Chen MF, Chen PY, Kuo JS, and Lee TJ. Role of Perivascular Adipose Tissue-Derived Methyl Palmitate in Vascular Tone Regulation and Pathogenesis of Hypertension. *Circulation*, 2011.

[249] Levy BI, Schiffrin EL, Mourad JJ, Agostini D, Vicaut E, Safar ME, and Struijker-Boudier HA. Impaired tissue perfusion: a pathology common to hypertension, obesity, and diabetes mellitus. *Circulation* 118: pp. 968–76, 2008.

[250] Lintschinger B, Balzer-Geldsetzer M, Baskaran T, Graier WF, Romanin C, Zhu MX, and Groschner K. Coassembly of Trp1 and Trp3 proteins generates diacylglycerol- and Ca2+-sensitive cation channels. *J. Biol. Chem.* 275: pp. 27799–805, 2000.

[251] Liu J, Hill MA, and Meininger GA. Mechanisms of myogenic enhancement by norepinephrine. *Am. J. Physiol.* 266: pp. H440–6, 1994.

[252] Liu Y, Bubolz AH, Mendoza S, Zhang DX, and Gutterman DD. H2O2 is the transferrable factor mediating flow-induced dilation in human coronary arterioles. *Circ. Res.* 108: pp. 566–73, 2011.

[253] Liu Y, Bubolz AH, Shi Y, Newman PJ, Newman DK, and Gutterman DD. Peroxynitrite reduces the endothelium-derived hyperpolarizing factor component of coronary flow-mediated dilation in PECAM-1-knockout mice. *Am. J. Physiol. Regul. Integr. Comp. Physiol.* 290: pp. R57–65, 2006.

[254] Lombard JH, and Duling BR. Multiple mechanisms of reactive hyperemia in arterioles of the hamster cheek pouch. *Am. J. Physiol.* 241: pp. H748–55, 1981.

[255] Lombard JH, and Stekiel WJ. Responses of cremasteric arterioles of spontaneously hypertensive rats to changes in extracellular K+ concentration. *Microcirculation* 2: pp. 355–62, 1995.

[256] Lott ME, Hogeman CS, Vickery L, Kunselman AR, Sinoway LI, and MacLean DA. Effects of dynamic exercise on mean blood velocity and muscle interstitial metabolite responses in humans. *Am. J. Physiol. Heart Circ. Physiol.* 281: pp. H1734–41, 2001.

[257] Loutzenhiser R, Bidani A, and Chilton L. Renal myogenic response: kinetic attributes and physiological role. *Circ. Res.* 90: pp. 1316–24, 2002.

[258] Loutzenhiser R, and Epstein M. Renal microvascular actions of calcium antagonists. *J. Am. Soc. Nephrol.* 1: pp. S3–12, 1990.

[259] Lundvall J. Tissue hyperosmolality as a mediator of vasodilatation and transcapillary fluid flux in exercising skeletal muscle. *Acta. Physiol. Scand. Suppl.* 379: pp. 1–142, 1972.

[260] Ma X, Qiu S, Luo J, Ma Y, Ngai CY, Shen B, Wong CO, Huang Y, and Yao X. Functional role of vanilloid transient receptor potential 4-canonical transient receptor potential 1 complex in flow-induced Ca2+ influx. *Arterioscler. Thromb. Vasc. Biol.* 30: pp. 851–8, 2010.

[261] MacLean DA, LaNoue KF, Gray KS, and Sinoway LI. Effects of hindlimb contraction on pressor and muscle interstitial metabolite responses in the cat. *J. Appl. Physiol.* 85: pp. 1583–92, 1998.

[262] Majesky MW, Dong XR, Hoglund V, Mahoney WM, Jr., and Daum G. The adventitia: a dynamic interface containing resident progenitor cells. *Arterioscler. Thromb. Vasc. Biol.* 31: pp. 1530–9, 2011.

[263] Mandala M, and Osol G. Physiological remodelling of the maternal uterine circulation during pregnancy. *Basic Clin. Pharmacol. Toxicol.* 110: pp. 12–8, 2012.

[264] Martinez-Lemus LA, Crow T, Davis MJ, and Meininger GA. alphavbeta3- and alpha5beta1-integrin blockade inhibits myogenic constriction of skeletal muscle resistance arterioles. *Am. J. Physiol. Heart. Circ. Physiol.* 289: pp. H322–9, 2005.

[265] Martinez-Lemus LA, and Galinanes EL. Matrix metalloproteinases and small artery remodeling. *Drug. Discov. Today Dis. Models* 8: pp. 21–8, 2011.

[266] Martinez-Lemus LA, Hill MA, Bolz SS, Pohl U, and Meininger GA. Acute mechano-adaptation of vascular smooth muscle cells in response to continuous arteriolar vasoconstriction: implications for functional remodeling. *Faseb. J.* 18: pp. 708–10, 2004.

[267] Martinez-Lemus LA, Hill MA, and Meininger GA. The plastic nature of the vascular wall: a continuum of remodeling events contributing to control of arteriolar diameter and structure. *Physiology (Bethesda)* 24: pp. 45–57, 2009.

[268] McCarron JG, Chalmers S, Bradley KN, MacMillan D, and Muir TC. Ca2+ microdomains in smooth muscle. *Cell Calcium* 40: pp. 461–93, 2006.

[269] McCarron JG, Crichton CA, Langton PD, MacKenzie A, and Smith GL. Myogenic contraction by modulation of voltage-dependent calcium currents in isolated rat cerebral arteries. *J. Physiol.* 498 (Pt 2): pp. 371–9, 1997.

[270] McCarron JG, and Halpern W. Potassium dilates rat cerebral arteries by two independent mechanisms. *Am. J. Physiol.* 259: pp. H902–8, 1990.

[271] McCarron JG, Osol G, and Halpern W. Myogenic responses are independent of the endothelium in rat pressurized posterior cerebral arteries. *Blood Vessels* 26: pp. 315–9, 1989.

[272] McGillivray-Anderson KM, and Faber JE. Effect of acidosis on contraction of microvascular smooth muscle by alpha 1- and alpha 2-adrenoceptors. Implications for neural and metabolic regulation. *Circ. Res.* 66: pp. 1643–57, 1990.

[273] Mederos y Schnitzler M, Storch U, Meibers S, Nurwakagari P, Breit A, Essin K, Gollasch M, and Gudermann T. Gq-coupled receptors as mechanosensors mediating myogenic vasoconstriction. *Embo. J.* 27: pp. 3092–103, 2008.

[274] Meininger GA, and Davis MJ. Cellular mechanisms involved in the vascular myogenic response. *Am. J. Physiol.* 263: pp. H647–59, 1992.

[275] Meininger GA, and Faber JE. Adrenergic facilitation of myogenic response in skeletal muscle arterioles. *Am. J. Physiol.* 260: pp. H1424–32, 1991.

[276] Meininger GA, Harris PD, and Joshua IG. Distributions of microvascular pressure in skeletal muscle of one-kidney, one clip, two-kidney, one clip, and deoxycorticosterone-salt hypertensive rats. *Hypertension* 6: pp. 27–34, 1984.

[277] Meininger GA, Mack CA, Fehr KL, and Bohlen HG. Myogenic vasoregulation overrides local metabolic control in resting rat skeletal muscle. *Circ. Res.* 60: pp. 861–70, 1987.

[278] Meininger GA, and Trzeciakowski JP. Combined effects of autoregulation and vasoconstrictors on hindquarters vascular resistance. *Am. J. Physiol.* 258: pp. H1032–41, 1990.

[279] Meininger GA, Zawieja DC, Falcone JC, Hill MA, and Davey JP. Calcium measurement in isolated arterioles during myogenic and agonist stimulation. *Am. J. Physiol.* 261: pp. H950–9, 1991.

[280] Mendoza SA, Fang J, Gutterman DD, Wilcox DA, Bubolz AH, Li R, Suzuki M, and Zhang DX. TRPV4-mediated endothelial Ca2+ influx and vasodilation in response to shear stress. *Am. J. Physiol. Heart. Circ. Physiol.* 298: pp. H466–76, 2010.

[281] Meyer B, Mortl D, Strecker K, Hulsmann M, Kulemann V, Neunteufl T, Pacher R, and Berger R. Flow-mediated vasodilation predicts outcome in patients with chronic heart failure: comparison with B-type natriuretic peptide. *J. Am. Coll. Cardiol.* 46: pp. 1011–8, 2005.

[282] Mo FM, and Ballard HJ. Adenosine output from dog gracilis muscle during systemic hypercapnia and/or amiloride-SITS infusion. *Am. J. Physiol.* 267: pp. H1243–9, 1994.

[283] Mochizuki S, Vink H, Hiramatsu O, Kajita T, Shigeto F, Spaan JA, and Kajiya F. Role of hyaluronic acid glycosaminoglycans in shear-induced endothelium-derived nitric oxide release. *Am. J. Physiol. Heart Circ. Physiol.* 285: pp. H722–6, 2003.

[284] Mogford JE, Davis GE, Platts SH, and Meininger GA. Vascular smooth muscle alpha v

beta 3 integrin mediates arteriolar vasodilation in response to RGD peptides. *Circ. Res.* 79: pp. 821–6, 1996.

[285] Moncada S, Gryglewski R, Bunting S, and Vane JR. An enzyme isolated from arteries transforms prostaglandin endoperoxides to an unstable substance that inhibits platelet aggregation. *Nature* 263: pp. 663–5, 1976.

[286] Moreno AP, and Lau AF. Gap junction channel gating modulated through protein phosphorylation. *Prog. Biophys. Mol. Biol.* 94: pp. 107–19, 2007.

[287] Moreno-Dominguez A, Colias O, El-Yazbi A, Walsh EJ, Hill MA, Walsh MP, and Cole WC. Calcium sensitization due to myosin light chain phosphatase inhibition and cytoskeletal reorganization in the myogenic response of skeletal muscle resistance arteries. *J. Physiol.*, 2012.

[288] Morff RJ, and Granger HJ. Autoregulation of blood flow within individual arterioles in the rat cremaster muscle. *Circ. Res.* 51: pp. 43–55, 1982.

[289] Morgan KG, and Gangopadhyay SS. Invited review: cross-bridge regulation by thin filament-associated proteins. *J. Appl. Physiol.* 91: pp. 953–62, 2001.

[290] Mori K, Nakaya Y, Sakamoto S, Hayabuchi Y, Matsuoka S, and Kuroda Y. Lactate-induced vascular relaxation in porcine coronary arteries is mediated by Ca2+-activated K+ channels. *J. Mol. Cell. Cardiol.* 30: pp. 349–56, 1998.

[291] Muller JM, Chilian WM, and Davis MJ. Integrin signaling transduces shear stress—dependent vasodilation of coronary arterioles. *Circ. Res.* 80: pp. 320–6, 1997.

[292] Mulvany MJ. Small artery remodelling in hypertension. *Basic Clin. Pharmacol. Toxicol.* 110: pp. 49–55, 2012.

[293] Mulvany MJ, and Halpern W. Mechanical properties of vascular smooth muscle cells in situ. *Nature* 260: pp. 617–9, 1976.

[294] Nakamura A, Hayashi K, Ozawa Y, Fujiwara K, Okubo K, Kanda T, Wakino S, and Saruta T. Vessel- and vasoconstrictor-dependent role of rho/rho-kinase in renal microvascular tone. *J. Vasc. Res.* 40: pp. 244–51, 2003.

[295] Narayanan D, Adebiyi A, and Jaggar JH. Inositol trisphosphate receptors in smooth muscle cells. *Am. J. Physiol. Heart Circ. Physiol.* 302: pp. H2190–210, 2012.

[296] Narayanan J, Imig M, Roman RJ, and Harder DR. Pressurization of isolated renal arteries increases inositol trisphosphate and diacylglycerol. *Am. J. Physiol.* 266: pp. H1840–5, 1994.

[297] Nausch LW, Bonev AD, Heppner TJ, Tallini Y, Kotlikoff MI, and Nelson MT. Sympathetic nerve stimulation induces local endothelial Ca2+ signals to oppose vasoconstriction of mouse mesenteric arteries. *Am. J. Physiol. Heart Circ. Physiol.* 302: pp. H594–602, 2012.

[298] Nellis SH, Liedtke AJ, and Whitesell L. Small coronary vessel pressure and diameter in an

intact beating rabbit heart using fixed-position and free-motion techniques. *Circ. Res.* 49: pp. 342–53, 1981.

[299] Nelson MT, Cheng H, Rubart M, Santana LF, Bonev AD, Knot HJ, and Lederer WJ. Relaxation of arterial smooth muscle by calcium sparks. *Science* 270: pp. 633–7, 1995.

[300] Nelson MT, and Quayle JM. Physiological roles and properties of potassium channels in arterial smooth muscle. *Am. J. Physiol.* 268: pp. C799–822, 1995.

[301] Nguyen TS, Winn HR, and Janigro D. ATP-sensitive potassium channels may participate in the coupling of neuronal activity and cerebrovascular tone. *Am. J. Physiol. Heart Circ. Physiol.* 278: pp. H878–85, 2000.

[302] Nijkamp FP, Flower RJ, Moncada S, and Vane JR. Partial purification of rabbit aorta contracting substance-releasing factor and inhibition of its activity by anti-inflammatory steroids. *Nature* 263: pp. 479–82, 1976.

[303] Nilius B, and Droogmans G. Ion channels and their functional role in vascular endothelium. *Physiol. Rev.* 81: pp. 1415–59, 2001.

[304] Norris CP, Barnes GE, Smith EE, and Granger HJ. Autoregulation of superior mesenteric flow in fasted and fed dogs. *Am. J. Physiol.* 237: pp. H174–7, 1979.

[305] Nurkiewicz TR, Porter DW, Hubbs AF, Stone S, Moseley AM, Cumpston JL, Goodwill AG, Frisbee SJ, Perrotta PL, Brock RW, Frisbee JC, Boegehold MA, Frazer DG, Chen BT, and Castranova V. Pulmonary particulate matter and systemic microvascular dysfunction. *Res. Rep. Health Eff. Inst.*: 3–48, 2011.

[306] Olesen SP, Clapham DE, and Davies PF. Haemodynamic shear stress activates a K+ current in vascular endothelial cells. *Nature* 331: pp. 168–170, 1988.

[307] Osawa M, Masuda M, Kusano K, and Fujiwara K. Evidence for a role of platelet endothelial cell adhesion molecule-1 in endothelial cell mechanosignal transduction: is it a mechanoresponsive molecule? *J. Cell. Biol.* 158: pp. 773–85, 2002.

[308] Osol G, Brekke JF, McElroy-Yaggy K, and Gokina NI. Myogenic tone, reactivity, and forced dilatation: a three-phase model of in vitro arterial myogenic behavior. *Am. J. Physiol. Heart Circ. Physiol.* 283: pp. H2260–7, 2002.

[309] Osol G, Laher I, and Cipolla M. Protein kinase C modulates basal myogenic tone in resistance arteries from the cerebral circulation. *Circ. Res.* 68: pp. 359–67, 1991.

[310] Osol G, Laher I, and Kelley M. Myogenic tone is coupled to phospholipase C and G protein activation in small cerebral arteries. *Am. J. Physiol.* 265: pp. H415–20, 1993.

[311] Osol G, and Mandala M. Maternal uterine vascular remodeling during pregnancy. *Physiology (Bethesda)* 24: pp. 58–71, 2009.

[312] Owens GK. Molecular control of vascular smooth muscle cell differentiation and phenotypic plasticity. *Novartis. Found. Symp.* 283: 174–191; discussion 191–173, 238–141, 2007.

[313] Pahakis MY, Kosky JR, Dull RO, and Tarbell JM. The role of endothelial glycocalyx components in mechanotransduction of fluid shear stress. *Biochem. Biophys. Res. Commun.* 355: pp. 228–33, 2007.

[314] Park KS, Kim Y, Lee YH, Earm YE, and Ho WK. Mechanosensitive cation channels in arterial smooth muscle cells are activated by diacylglycerol and inhibited by phospholipase C inhibitor. *Circ. Res.* 93: pp. 557–64, 2003.

[315] Perry MA, and Granger DN. Role of CD11/CD18 in shear rate-dependent leukocyte-endothelial cell interactions in cat mesenteric venules. *J. Clin. Invest.* 87: pp. 1798–804, 1991.

[316] Poburko D, Fameli N, Kuo KH, and van Breemen C. Ca2+ signaling in smooth muscle: TRPC6, NCX and LNats in nanodomains. *Channels (Austin)* 2: 10–2, 2008.

[317] Pohl U, Herlan K, Huang A, and Bassenge E. EDRF-mediated shear-induced dilation opposes myogenic vasoconstriction in small rabbit arteries. *Am. J. Physiol.* 261: pp. H2016–23, 1991.

[318] Porter M, Evans MC, Miner AS, Berg KM, Ward KR, and Ratz PH. Convergence of Ca2+-desensitizing mechanisms activated by forskolin and phenylephrine pretreatment, but not 8-bromo-cGMP. *Am. J. Physiol. Cell. Physiol.* 290: pp. C1552–9, 2006.

[319] Pucovsky V, Harhun MI, Povstyan OV, Gordienko DV, Moss RF, and Bolton TB. Close relation of arterial ICC-like cells to the contractile phenotype of vascular smooth muscle cell. *J. Cell. Mol. Med.* 11: pp. 764–75, 2007.

[320] Puro DG. Retinovascular physiology and pathophysiology: new experimental approach/ new insights. *Prog. Retin. Eye. Res.* 31: pp. 258–70, 2012.

[321] Quayle JM, Nelson MT, and Standen NB. ATP-sensitive and inwardly rectifying potassium channels in smooth muscle. *Physiol. Rev.* 77: pp. 1165–232, 1997.

[322] Raina H, Ella SR, and Hill MA. Decreased activity of the smooth muscle Na+/Ca2+ exchanger impairs arteriolar myogenic reactivity. *J. Physiol.* 586: pp. 1669–81, 2008.

[323] Ray CJ, and Marshall JM. Elucidation in the rat of the role of adenosine and A2A-receptors in the hyperaemia of twitch and tetanic contractions. *J. Physiol.* 587: pp. 1565–78, 2009.

[324] Riddle DR, Sonntag WE, and Lichtenwalner RJ. Microvascular plasticity in aging. *Ageing Res. Rev.* 2: pp. 149–68, 2003.

[325] Rivers RJ. Remote effects of pressure changes in arterioles. *Am. J. Physiol.* 268: pp. H1379–82, 1995.

[326] Rizzo V, Morton C, DePaola N, Schnitzer JE, and Davies PF. Recruitment of endothelial caveolae into mechanotransduction pathways by flow conditioning in vitro. *Am. J. Physiol. Heart Circ. Physiol.* 285: pp. H1720–9, 2003.

[327] Rizzo V, Sung A, Oh P, and Schnitzer JE. Rapid mechanotransduction in situ at the luminal cell surface of vascular endothelium and its caveolae. *J. Biol. Chem.* 273: pp. 26323–9, 1998.

[328] Rizzoni D, Porteri E, Boari GE, De Ciuceis C, Sleiman I, Muiesan ML, Castellano M, Miclini M, and Agabiti-Rosei E. Prognostic significance of small-artery structure in hypertension. *Circulation* 108: pp. 2230–5, 2003.

[329] Rizzoni D, Porteri E, Guelfi D, Muiesan ML, Valentini U, Cimino A, Girelli A, Rodella L, Bianchi R, Sleiman I, and Rosei EA. Structural alterations in subcutaneous small arteries of normotensive and hypertensive patients with non-insulin-dependent diabetes mellitus. *Circulation* 103: pp. 1238–44, 2001.

[330] Roman R. In: *Handbook of Physiology, The Cardiovascular System, Vascular Smooth Muscle.* San Diego: Academic Press, 2008.

[331] Roman RJ, Renic M, Dunn KM, Takeuchi K, and Hacein-Bey L. Evidence that 20-HETE contributes to the development of acute and delayed cerebral vasospasm. *Neurol. Res.* 28: pp. 738–49, 2006.

[332] Rosendal L, Blangsted AK, Kristiansen J, Sogaard K, Langberg H, Sjogaard G, and Kjaer M. Interstitial muscle lactate, pyruvate and potassium dynamics in the trapezius muscle during repetitive low-force arm movements, measured with microdialysis. *Acta. Physiol. Scand.* 182: pp. 379–88, 2004.

[333] Roy CS, and Sherrington CS. On the Regulation of the Blood-supply of the Brain. *J. Physiol.* 11: pp. 85–158, 1890.

[334] Rubio R, and Berne RM. Release of adenosine by the normal myocardium in dogs and its relationship to the regulation of coronary resistance. *Circ. Res.* 25: pp. 407–15, 1969.

[335] Rubio R, Berne RM, and Katori M. Release of adenosine in reactive hyperemia of the dog heart. *Am. J. Physiol.* 216: pp. 56–62, 1969.

[336] Sanders KM. A case for interstitial cells of Cajal as pacemakers and mediators of neurotransmission in the gastrointestinal tract. *Gastroenterology* 111: pp. 492–515, 1996.

[337] Sanders KM, Ordog T, Koh SD, and Ward SM. A Novel Pacemaker Mechanism Drives Gastrointestinal Rhythmicity. *News. Physiol. Sci.* 15: pp. 291–8, 2000.

[338] Sandow SL. Factors, fiction and endothelium-derived hyperpolarizing factor. *Clin. Exp. Pharmacol. Physiol.* 31: pp. 563–70, 2004.

[339] Sandow SL, Haddock RE, Hill CE, Chadha PS, Kerr PM, Welsh DG, and Plane F. What's where and why at a vascular myoendothelial microdomain signalling complex. *Clin. Exp. Pharmacol. Physiol.* 36: pp. 67–76, 2009.

[340] Sandow SL, Neylon CB, Chen MX, and Garland CJ. Spatial separation of endothelial small- and intermediate-conductance calcium-activated potassium channels (K(Ca)) and connexins: possible relationship to vasodilator function? *J. Anat.* 209: pp. 689–98, 2006.

[341] Sandow SL, Tare M, Coleman HA, Hill CE, and Parkington HC. Involvement of myoendothelial gap junctions in the actions of endothelium-derived hyperpolarizing factor. *Circ. Res.* 90: pp. 1108–13, 2002.

[342] Sato A, Terata K, Miura H, Toyama K, Loberiza FR, Jr., Hatoum OA, Saito T, Sakuma I, and Gutterman DD. Mechanism of vasodilation to adenosine in coronary arterioles from patients with heart disease. *Am. J. Physiol. Heart. Circ. Physiol.* 288: pp. H1633–40, 2005.

[343] Schrage WG, Wilkins BW, Dean VL, Scott JP, Henry NK, Wylam ME, and Joyner MJ. Exercise hyperemia and vasoconstrictor responses in humans with cystic fibrosis. *J. Appl. Physiol.* 99: pp. 1866–71, 2005.

[344] Schretzenmayr A. Uber kreis laufregulatorische vorgange an den grobem arterien bei der muskelarbeit. *Pflugers Archives European. Journal of Physiology* 232: pp. 743–8, 1933.

[345] Schubert R, Kalentchuk VU, and Krien U. Rho kinase inhibition partly weakens myogenic reactivity in rat small arteries by changing calcium sensitivity. *Am. J. Physiol. Heart. Circ. Physiol.* 283: pp. H2288–95, 2002.

[346] Schubert R, Lidington D, and Bolz SS. The emerging role of Ca2+ sensitivity regulation in promoting myogenic vasoconstriction. *Cardiovasc. Res.* 77: pp. 8–18, 2008.

[347] Scott JB, Hardin RA, and Haddy FJ. Pressure-flow relationships in the coronary vascular bed of the dog. *Am. J. Physiol.* 199: pp. 765–9, 1960.

[348] Segal SS, and Duling BR. Flow control among microvessels coordinated by intercellular conduction. *Science* 234: pp. 868–70, 1986.

[349] Segal SS, and Jacobs TL. Role for endothelial cell conduction in ascending vasodilatation and exercise hyperaemia in hamster skeletal muscle. *J. Physiol.* 536: pp. 937–46, 2001.

[350] Selkurt EE. The relation of renal blood flow to effective arterial pressure in the intact kidney of the dog. *Am. J. Physiol.* 147: pp. 537–49, 1946.

[351] Setoguchi M, Ohya Y, Abe I, and Fujishima M. Stretch-activated whole-cell currents in smooth muscle cells from mesenteric resistance artery of guinea-pig. *J. Physiol.* 501 (Pt 2): pp. 343–53, 1997.

[352] Shapiro HM, Stromberg DD, Lee DR, and Wiederhielm CA. Dynamic pressures in the pial arterial microcirculation. *Am. J. Physiol.* 221: pp. 279–83, 1971.

[353] Shimokawa H, Yasutake H, Fujii K, Owada MK, Nakaike R, Fukumoto Y, Takayanagi T, Nagao T, Egashira K, Fujishima M, and Takeshita A. The importance of the hyperpolarizing mechanism increases as the vessel size decreases in endothelium-dependent relaxations in rat mesenteric circulation. *J. Cardiovasc. Pharmacol.* 28: pp. 703–11, 1996.

[354] Shiraishi M, Wang X, Walsh MP, Kargacin G, Loutzenhiser K, and Loutzenhiser R. Myosin heavy chain expression in renal afferent and efferent arterioles: relationship to contractile kinetics and function. *FASEB. J.* 17: pp. 2284–6, 2003.

[355] Shirao S, Fujisawa H, Kudo A, Kurokawa T, Yoneda H, Kunitsugu I, Ogasawara K, Soma M, Kobayashi S, Ogawa A, and Suzuki M. Inhibitory effects of eicosapentaenoic acid on chronic cerebral vasospasm after subarachnoid hemorrhage: possible involvement of a sphingosylphosphorylcholine-rho-kinase pathway. *Cerebrovasc. Dis.* 26: pp. 30–7, 2008.

[356] Simmons GH, Padilla J, and Laughlin MH. Heterogeneity of endothelial cell phenotype within and amongst conduit vessels of the swine vasculature. *Exp. Physiol.*, 2012.

[357] Slish DF, Welsh DG, and Brayden JE. Diacylglycerol and protein kinase C activate cation channels involved in myogenic tone. *Am. J. Physiol. Heart. Circ. Physiol.* 283: pp. H2196–201, 2002.

[358] Smiesko V, Lang DJ, and Johnson PC. Dilator response of rat mesenteric arcading arterioles to increased blood flow velocity. *Am. J. Physiol.* 257: pp. H1958–65, 1989.

[359] Smolock EM, Trappanese DM, Chang S, Wang T, Titchenell P, and Moreland RS. siRNA-mediated knockdown of h-caldesmon in vascular smooth muscle. *Am. J. Physiol. Heart. Circ. Physiol.* 297: pp. H1930–9, 2009.

[360] Sobey CG. Potassium channel function in vascular disease. *Arterioscler. Thromb. Vasc. Biol.* 21: pp. 28–38, 2001.

[361] Somlyo AP, and Somlyo AV. Ca2+ sensitivity of smooth muscle and nonmuscle myosin II: modulated by G proteins, kinases, and myosin phosphatase. *Physiol. Rev.* 83: pp. 1325–58, 2003.

[362] Somlyo AP, and Somlyo AV. Signal transduction and regulation in smooth muscle. *Nature* 372: pp. 231–6, 1994.

[363] Sonkusare SK, Bonev AD, Ledoux J, Liedtke W, Kotlikoff MI, Heppner TJ, Hill-Eubanks DC, and Nelson MT. Elementary Ca2+ signals through endothelial TRPV4 channels regulate vascular function. *Science* 336: pp. 597–601, 2012.

[364] Spassova MA, Hewavitharana T, Xu W, Soboloff J, and Gill DL. A common mechanism underlies stretch activation and receptor activation of TRPC6 channels. *Proc. Natl. Acad. Sci. U S A* 103: pp. 16586–91, 2006.

[365] Sprague RS, Bowles EA, Achilleus D, and Ellsworth ML. Erythrocytes as controllers of perfusion distribution in the microvasculature of skeletal muscle. *Acta. Physiol. (Oxf.)* 202: pp. 285–92, 2011.

[366] Sprague RS, Ellsworth ML, Stephenson AH, and Lonigro AJ. ATP: the red blood cell link to NO and local control of the pulmonary circulation. *Am. J. Physiol.* 271: pp. H2717–22, 1996.

[367] Stainsby WN, and Renkin EM. Autoregulation of blood flow in peripheral vascular beds. *Am. J. Cardiol.* 8: pp. 741–7, 1961.

[368] Stansberry KB, Hill MA, Shapiro SA, McNitt PM, Bhatt BA, and Vinik AI. Impairment of peripheral blood flow responses in diabetes resembles an enhanced aging effect. *Diabetes Care* 20: pp. 1711–6, 1997.

[369] Stansberry KB, Shapiro SA, Hill MA, McNitt PM, Meyer MD, and Vinik AI. Impaired peripheral vasomotion in diabetes. *Diabetes Care* 19: pp. 715–21, 1996.

[370] Starling EH. On the Absorption of Fluids from the Connective Tissue Spaces. *J. Physiol.* 19: pp. 312–26, 1896.

[371]   Steenbergen JM, and Bohlen HG. Sodium hyperosmolarity of intestinal lymph causes arteriolar vasodilation in part mediated by EDRF. *Am. J. Physiol.* 265: pp. H323–8, 1993.

[372]   Stoner L, Erickson ML, Young JM, Fryer S, Sabatier MJ, Faulkner J, Lambrick DM, and McCully KK. There's more to flow-mediated dilation than nitric oxide. *J. Atheroscler. Thromb.* 2012.

[373]   Storch U, Mederos y Schnitzler M, and Gudermann T. G protein-mediated stretch reception. *Am. J. Physiol. Heart. Circ. Physiol.* 302: pp. H1241–9, 2012.

[374]   Stratman AN, and Davis GE. Endothelial cell-pericyte interactions stimulate basement membrane matrix assembly: influence on vascular tube remodeling, maturation, and stabilization. *Microsc. Microanal.* 18: pp. 68–80, 2012.

[375]   Stratman AN, Malotte KM, Mahan RD, Davis MJ, and Davis GE. Pericyte recruitment during vasculogenic tube assembly stimulates endothelial basement membrane matrix formation. *Blood* 114: pp. 5091–101, 2009.

[376]   Stromberg DD, and Fox JR. Pressures in the pial arterial microcirculation of the cat during changes in systemic arterial blood pressure. *Circ. Res.* 31: pp. 229–39, 1972.

[377]   Sun D, Messina EJ, Kaley G, and Koller A. Characteristics and origin of myogenic response in isolated mesenteric arterioles. *Am. J. Physiol.* 263: pp. H1486–91, 1992.

[378]   Sun Z, Martinez-Lemus LA, Hill MA, and Meininger GA. Extracellular matrix-specific focal adhesions in vascular smooth muscle produce mechanically active adhesion sites. *Am. J. Physiol. Cell. Physiol.* 295: pp. C268–78, 2008.

[379]   Sweeney TE, and Sarelius IH. Spatial heterogeneity in striated muscle arteriolar tone, cell flow, and capillarity. *Am. J. Physiol.* 259: pp. H124–36, 1990.

[380]   Tabouillot T, Muddana HS, and Butler PJ. Endothelial Cell Membrane Sensitivity to Shear Stress is Lipid Domain Dependent. *Cell. Mol. Bioeng.* 4: pp. 169–81, 2011.

[381]   Takeya K, Loutzenhiser K, Shiraishi M, Loutzenhiser R, and Walsh MP. A highly sensitive technique to measure myosin regulatory light chain phosphorylation: the first quantification in renal arterioles. *Am. J. Physiol. Renal. Physiol.* 294: pp. F1487–92, 2008.

[382]   Tallini YN, Brekke JF, Shui B, Doran R, Hwang SM, Nakai J, Salama G, Segal SS, and Kotlikoff MI. Propagated endothelial Ca2+ waves and arteriolar dilation in vivo: measurements in Cx40BAC GCaMP2 transgenic mice. *Circ. Res.* 101: pp. 1300–9, 2007.

[383]   Tateishi J, and Faber JE. Inhibition of arteriole alpha 2- but not alpha 1-adrenoceptor constriction by acidosis and hypoxia in vitro. *Am. J. Physiol.* 268: pp. H2068–76, 1995.

[384]   Thomas CE, Ott CE, Bell PD, Knox FG, and Navar LG. Glomerular filtration dynamics during renal vasodilation with acetylcholine in the dog. *Am. J. Physiol.* 244: pp. F606–11, 1983.

[385]   To WJ, O'Brien VP, Banerjee A, Gutierrez AN, Li J, Chen PC, and Cheung AT. Real-time

studies of hypertension using non-mydriatic fundus photography and computer-assisted intravital microscopy. *Clin. Hemorheol. Microcirc.*, 2012.

[386] Tooke JE, Lins PE, Ostergren J, and Fagrell B. Skin microvascular autoregulatory responses in type I diabetes: the influence of duration and control. *Int. J. Microcirc. Clin. Exp.* 4: pp. 249–56, 1985.

[387] Tooke JE, Ostergren J, and Fagrell B. Synchronous assessment of human skin microcirculation by laser Doppler flowmetry and dynamic capillaroscopy. *Int. J. Microcirc. Clin. Exp.* 2: pp. 277–84, 1983.

[388] Tran CH, Taylor MS, Plane F, Nagaraja S, Tsoukias NM, Solodushko V, Vigmond EJ, Furstenhaupt T, Brigdan M, and Welsh DG. Endothelial Ca2+ wavelets and the induction of myoendothelial feedback. *Am. J. Physiol. Cell. Physiol.* 302: pp. C1226–42, 2012.

[389] Tuma RF. The cerebral microcirculation. In: *Handbook of Physiology, Microcirculation* (2nd ed.), edited by Tuma RF, Duran WN, Ley K. San Diego: Academic Press, 2008, pp. 485–520.

[390] Tuma RF, Lindbom L, and Arfors KE. Dependence of reactive hyperemia in skeletal muscle on oxygen tension. *Am. J. Physiol.* 233: pp. H289–94, 1977.

[391] Tumelty J, Scholfield N, Stewart M, Curtis T, and McGeown G. Ca2+-sparks constitute elementary building blocks for global Ca2+-signals in myocytes of retinal arterioles. *Cell Calcium* 41: pp. 451–66, 2007.

[392] Tune JD, Gorman MW, and Feigl EO. Matching coronary blood flow to myocardial oxygen consumption. *J. Appl. Physiol.* 97: pp. 404–15, 2004.

[393] Tune JD, Richmond KN, Gorman MW, Olsson RA, and Feigl EO. Adenosine is not responsible for local metabolic control of coronary blood flow in dogs during exercise. *Am. J. Physiol. Heart. Circ. Physiol.* 278: pp. H74–84, 2000.

[394] Uchida E, and Bohr DF. Myogenic tone in isolated perfused resistance vessels from rats. *Am. J. Physiol.* 216: pp. 1343–50, 1969.

[395] Uchida E, and Bohr DF. Myogenic tone in isolated perfused vessels. Occurrence among vascular beds and along vascular trees. *Circ. Res.* 25: pp. 549–55, 1969.

[396] Unger VM, Kumar NM, Gilula NB, and Yeager M. Three-dimensional structure of a recombinant gap junction membrane channel. *Science* 283: pp. 1176–80, 1999.

[397] Ungvari Z, and Koller A. Selected contribution: NO released to flow reduces myogenic tone of skeletal muscle arterioles by decreasing smooth muscle Ca(2+) sensitivity. *J. Appl. Physiol.* 91: 522–7; discussion 504–25, 2001.

[398] VanBavel E, van der Meulen ET, and Spaan JA. Role of Rho-associated protein kinase in tone and calcium sensitivity of cannulated rat mesenteric small arteries. *Exp. Physiol.* 86: pp. 585–92, 2001.

[399] Vanhoutte PM, Shimokawa H, Tang EH, and Feletou M. Endothelial dysfunction and vascular disease. *Acta. Physiol. (Oxf.)* 196: pp. 193–222, 2009.

[400] Vennekens R. Emerging concepts for the role of TRP channels in the cardiovascular system. *J. Physiol.* 589: pp. 1527–34, 2011.

[401] von Anrep G. On local vascular reactions and their interpretation. *J. Physiol.* 45: pp. 318–27, 1912.

[402] Wahl M, Kuschinsky W, Bosse O, and Thurau K. Dependency of pial arterial and arteriolar diameter on perivascular osmolarity in the cat. A microapplication study. *Circ. Res.* 32: pp. 162–9, 1973.

[403] Walmsley JG, Gore RW, Dacey RG, Jr., Damon DN, and Duling BR. Quantitative morphology of arterioles from the hamster cheek pouch related to mechanical analysis. *Microvasc. Res.* 24: pp. 249–71, 1982.

[404] Ward SM. Interstitial cells of Cajal in enteric neurotransmission. *Gut.* 47 Suppl 4: iv40–3; discussion iv52, 2000.

[405] Watts SW, Kanagy NL, and Lombard JH. Receptor-mediated events in the microcirculation. In: *Handbook of Physiology, Microcirculation*, edited by Tuma RF, Duran WN, Ley K. San Diego: Academic Press, 2008, p. 285–347.

[406] Wayland H, and Johnson PC. Erythrocyte velocity measurement in microvessels by a two-slit photometric method. *J. Appl. Physiol.* 22: pp. 333–7, 1967.

[407] Wayland H, and Johnson PC. Optical scanning photometry for microcirculatory studies. *Bibl. Anat.* 10: pp. 564–70, 1969.

[408] Webb RL, and Nicoll PA. Circulatory flow pattern in the bat's wing. *Anat. Rec.* 97: p. 431, 1947.

[409] Wei AD, Gutman GA, Aldrich R, Chandy KG, Grissmer S, and Wulff H. International Union of Pharmacology. LII. Nomenclature and molecular relationships of calcium-activated potassium channels. *Pharmacol. Rev.* 57: pp. 463–72, 2005.

[410] Wei EP, Kontos HA, and Patterson JL, Jr. Dependence of pial arteriolar response to hypercapnia on vessel size. *Am. J. Physiol.* 238: pp. 697–703, 1980.

[411] Weinbaum S, Zhang X, Han Y, Vink H, and Cowin SC. Mechanotransduction and flow across the endothelial glycocalyx. *Proc. Natl. Acad. Sci. U S A* 100: pp. 7988–95, 2003.

[412] Wellman GC, Nathan DJ, Saundry CM, Perez G, Bonev AD, Penar PL, Tranmer BI, and Nelson MT. Ca2+ sparks and their function in human cerebral arteries. *Stroke* 33: pp. 802–8, 2002.

[413] Wellman GC, Santana LF, Bonev AD, and Nelson MT. Role of phospholamban in the modulation of arterial Ca(2+) sparks and Ca(2+)-activated K(+) channels by cAMP. *Am. J. Physiol. Cell. Physiol.* 281: pp. C1029–37, 2001.

[414] Welsh DG, Morielli AD, Nelson MT, and Brayden JE. Transient receptor potential channels regulate myogenic tone of resistance arteries. *Circ. Res.* 90: pp. 248–50, 2002.

[415] Welsh DG, Nelson MT, Eckman DM, and Brayden JE. Swelling-activated cation channels mediate depolarization of rat cerebrovascular smooth muscle by hyposmolarity and intravascular pressure. *J. Physiol.* 527 (Pt 1): pp. 139–48, 2000.

[416] Westcott EB, and Jackson WF. Heterogeneous function of ryanodine receptors, but not IP3 receptors, in hamster cremaster muscle feed arteries and arterioles. *Am. J. Physiol. Heart. Circ. Physiol.* 300: pp. H1616–30, 2011.

[417] Wiederhielm CA, Bouskela E, Heald R, and Black L. A method for varying arterial and venous pressures in intact, unanesthetized mammals. *Microvasc. Res.* 18: pp. 124–8, 1979.

[418] Wiederhielm CA, Woodbury JW, Kirk S, and Rushmer RF. Pulsatile Pressures in the Microcirculation of Frog's Mesentery. *Am. J. Physiol.* 207: pp. 173–6, 1964.

[419] Wilson DF, Erecinska M, Drown C, and Silver IA. Effect of oxygen tension on cellular energetics. *Am. J. Physiol.* 233: pp. C135–40, 1977.

[420] Wolfle SE, Schmidt VJ, Hoepfl B, Gebert A, Alcolea S, Gros D, and de Wit C. Connexin45 cannot replace the function of connexin40 in conducting endothelium-dependent dilations along arterioles. *Circ. Res.* 101: pp. 1292–9, 2007.

[421] Wu X, Davis GE, Meininger GA, Wilson E, and Davis MJ. Regulation of the L-type calcium channel by alpha 5beta 1 integrin requires signaling between focal adhesion proteins. *J. Biol. Chem.* 276: pp. 30285–92, 2001.

[422] Wu X, Mogford JE, Platts SH, Davis GE, Meininger GA, and Davis MJ. Modulation of calcium current in arteriolar smooth muscle by alphav beta3 and alpha5 beta1 integrin ligands. *J. Cell. Biol.* 143: pp. 241–52, 1998.

[423] Wu X, Yang Y, Gui P, Sohma Y, Meininger GA, Davis GE, Braun AP, and Davis MJ. Potentiation of large conductance, Ca2+-activated K+ (BK) channels by alpha5beta1 integrin activation in arteriolar smooth muscle. *J. Physiol.* 586: pp. 1699–713, 2008.

[424] Yamanishi S, Katsumura K, Kobayashi T, and Puro DG. Extracellular lactate as a dynamic vasoactive signal in the rat retinal microvasculature. *Am. J. Physiol. Heart. Circ. Physiol.* 290: pp. H925–34, 2006.

[425] Yang Y, Ella SR, Braun AP, Korhuis RJ, Davis MJ, and Hill MA. Manipulation of arteriolar BKCa using subunit directed siRNA (Abstract). *FASEB. J.*, 2010.

[426] Yang Y, Murphy TV, Ella SR, Grayson TH, Haddock R, Hwang YT, Braun AP, Peichun G, Korthuis RJ, Davis MJ, and Hill MA. Heterogeneity in function of small artery smooth muscle BKCa: involvement of the beta1-subunit. *J. Physiol.* 587: pp. 3025–44, 2009.

[427] Yu J, Bergaya S, Murata T, Alp IF, Bauer MP, Lin MI, Drab M, Kurzchalia TV, Stan RV,

and Sessa WC. Direct evidence for the role of caveolin-1 and caveolae in mechanotransduction and remodeling of blood vessels. *J. Clin. Invest.* 116: pp. 1284–91, 2006.

[428]   Zaritsky JJ, Eckman DM, Wellman GC, Nelson MT, and Schwarz TL. Targeted disruption of Kir2.1 and Kir2.2 genes reveals the essential role of the inwardly rectifying K(+) current in K(+)-mediated vasodilation. *Circ. Res.* 87: pp. 160–6, 2000.

[429]   Zhang DX, and Gutterman DD. Transient receptor potential channel activation and endothelium-dependent dilation in the systemic circulation. *J. Cardiovasc. Pharmacol.* 57: pp. 133–9, 2011.

[430]   Zhang J, Ren C, Chen L, Navedo MF, Antos LK, Kinsey SP, Iwamoto T, Philipson KD, Kotlikoff MI, Santana LF, Wier WG, Matteson DR, and Blaustein MP. Knockout of Na+/Ca2+ exchanger in smooth muscle attenuates vasoconstriction and L-type Ca2+ channel current and lowers blood pressure. *Am. J. Physiol. Heart. Circ. Physiol.* 298: pp. H1472–83, 2010.

[431]   Zhao G, Neeb ZP, Leo MD, Pachuau J, Adebiyi A, Ouyang K, Chen J, and Jaggar JH. Type 1 IP3 receptors activate BKCa channels via local molecular coupling in arterial smooth muscle cells. *J. Gen. Physiol.* 136: pp. 283–91, 2010.

[432]   Zhao G, Zhao Y, Pan B, Liu J, Huang X, Zhang X, Cao C, Hou N, Wu C, Zhao KS, and Cheng H. Hypersensitivity of BKCa to Ca2+ sparks underlies hyporeactivity of arterial smooth muscle in shock. *Circ. Res.* 101: pp. 493–502, 2007.

[433]   Zieba BJ, Artamonov MV, Jin L, Momotani K, Ho R, Franke AS, Neppl RL, Stevenson AS, Khromov AS, Chrzanowska-Wodnicka M, and Somlyo AV. The cAMP-responsive Rap1 guanine nucleotide exchange factor, Epac, induces smooth muscle relaxation by down-regulation of RhoA activity. *J. Biol. Chem.* 286: pp. 16681–92, 2011.

[434]   Zou H, Ratz PH, and Hill MA. Role of myosin phosphorylation and [Ca2+]i in myogenic reactivity and arteriolar tone. *Am. J. Physiol.* 269: pp. H1590–6, 1995.

[435]   Zou H, Ratz PH, and Hill MA. Temporal aspects of Ca(2+) and myosin phosphorylation during myogenic and norepinephrine-induced arteriolar constriction. *J. Vasc. Res.* 37: pp. 556–67, 2000.

Author Biographies

Dr. Michael Hill is a Professor of Medical Pharmacology and Physiology in the School of Medicine and Associate Director of the Dalton Cardiovascular Research Center, University of Missouri, Columbia, Missouri. He received his Ph.D. from the University of Melbourne before undertaking postdoctoral training at Texas A&M University. Prior to his position at the University of Missouri, Dr. Hill held academic positions at Eastern Virginia Medical School (Norfolk, VA), RMIT University (Melbourne, Australia) and the University of New South Wales (Sydney, Australia).

Dr. Hill's research focuses on myogenic mechanisms underlying the control of arteriolar diameter. Specifically, his work has centered on Ca^{2+} signaling, kinase activation and the role of ion channels. Recent work has also extended to the interaction between arteriolar wall structure and function. Research in his laboratory has been funded by the National Institutes of Health; National Health and Medical Research Council (Australia); American Heart Association; National Heart Foundation of Australia and the Juvenile Diabetes Foundation.

Dr. Michael Davis is a Professor of Medical Pharmacology and Physiology in the School of Medicine at the University of Missouri, Columbia, Missouri. Dr. Davis obtained his Ph.D. at the University of Nebraska Medical Center followed by postdoctoral training at the University of Arizona. Prior to his position at the University of Missouri Dr. Davis spent 20 years at Texas A&M University as Assistant, Associate and Professor of Medical Physiology.

Dr. Davis' research focuses on mechanotransduction by blood and lymphatic vessels. Research in his laboratory has been continuously funded by the National Institutes of Health since 1986.